D0874344

RADAR, HULA HOOPS, AND PLAYFUL PIGS

RADAR, HULA HOOPS, AND PLAYFUL PIGS

67 Digestible commentaries on the Fascinating Chemistry of Everyday Life

DR. JOE SCHWARCZ

W. H. Freeman and Company
New York

The publication of *Radar, Hula Hoops, and Playful Pigs* has been
generously supported by the Canada Council, the Ontario Arts Council,
and the Government of Canada through the Book Publishing Industry
Development Program.

Cover design by Guylaine Regimbald.
Cover Illustration ©David La Fleur/SIS.
Interior design by Yolande Martel.
Interior cartoons by Brian Gable.
Author photo by Georgio Campana.

Cataloging-in-publication data available from the Library of Congress

Published in the United States in 2001 by W. H. Freeman and Company,
41 Madison Avenue, New York, New York 10010.

Printed in the United States of America

Second US printing 2002.

CONTENTS

DOWN THE HATCH

CHEMICAL CRIMES

THE BOTTOM LINE

INTRODUCTION

A QUEST FOR THE RIGHT CHEMISTRY

When I was about 10 years old, I was invited to a birthday party. It turned out to be a life-changing event. We were entertained by a teenaged magician with the usual repertoire; his tricks were all forgettable — except for one. At a point in the less-than-bedazzling performance, the bored prestidigitator picked up three different-colored ropes and proceeded to tie them together. He then rolled them up in his hands and reached into his pocket for an invisible "magic chemical," which he pretended to sprinkle over the ropes. Wouldn't you know it, when he unraveled the ropes, the knots were gone and the three ropes had fused into one long rope!

I think even at that young age I realized that I had witnessed sleight of hand and not chemical magic. But I remember instantly wondering why this conjurer had chosen to amaze us with "chemicals" instead of the usual abracadabra or hocus-pocus. I didn't know anything about chemistry in those days and had no idea what chemicals were. Why had he associated chemicals with magic? I decided to find out. And am I ever glad I did. I have been enthralled with the magic of chemistry ever since that fateful birthday party.

My local library, as it turned out, had several books on chemical magic. Within weeks, I had learned how to change water into "wine," prepare invisible inks, and make self-lighting candles. It was fun. In fact, I still enjoy entertaining children of all ages with "chemical" magic shows. But then, as I read more and more, I discovered that the real magic of chemistry lay elsewhere. It didn't have anything to do with changing solutions from one color to another or with producing puffs and bangs. The real magic was to be found in chemistry's ability to unravel the mysteries of life.

For me, an understanding of molecules and their reactions demystified the workings of the world and perhaps, even more significantly, demonstrated the intimate link between the quality of daily life and chemical knowledge. The fragrance of a rose, the taste of an apple, the color of a carrot, the sting of a bee, the misery of an allergy, the tarnishing of silver, the pleasures of chocolate, and the secrets of love all surrendered their mysteries to an understanding of molecular behavior. Delving into chemistry cast light upon the effects of medications, the role of cosmetics, the principles of nutrition, the risks of toxins, the effectiveness of cleaning agents, the dangers of pollutants, and the horrors of chemical warfare. It became clear to me that you couldn't possibly navigate through life properly without an understanding of chemistry because basically we are all practicing chemists. We brew coffee, we cook, we paint, we wash, we eat, we have sex. We are constantly chemically challenged. We have to make decisions about which toothpaste, which shampoo, which detergent, and which vitamin supplement to use. We are obliged, therefore, not to fear chemicals but to learn about them.

But even this prospect scares a lot of people. Just think of the word "chemistry." What comes to your mind? Difficult? Boring? Dangerous? Polluting? Cancer-causing? Explosive?

Smelly? Unfortunately, when I've asked this question before, I've heard all these answers. The association is almost always unfavorable. Occasionally someone will halfheartedly murmur "Bunsen burner" or "periodic table," but the adjectives "interesting," "exciting," "amazing" almost never roll off the tongue. In a recent survey of students about to embark on their first chemistry course at a large American university, one student anticipated that the experience would be "antiseptically arrogant." I'm not exactly sure what this means, but somehow I don't think he was looking forward to an agreeable experience.

It is my contention that a little intellectual dip into the vast ocean that is chemistry can not only be useful but pleasurable as well. I have not attempted to produce a chemistry textbook here; there are numerous excellent ones around. Nor have I tried to engage in a comprehensive treatment of medications, food additives, cosmetics, or cleaning agents, although I hope you will also find much useful information on these topics here. Rather, I have attempted to build a framework for rational, scientific thought through a series of entertaining glimpses into the nooks and crannies of the world of science.

To me the phrase "the right chemistry" actually has two

connotations. The obvious one concerns knowing something about how molecules can be expected to behave. But I also look on the expression as a metaphor for a good mix. Couples and athletic teams can have the right chemistry, and so can concepts and ideas. I hope that this book reflects both of these notions, and that after darting with me through some of the following chemical escapades you'll appreciate why I'm so glad that I was invited to that birthday party so many years ago where a teenaged conjurer and his "magic chemical" roped me into the quest for "the right chemistry."

EVANGELISTS, MAD SCIENTISTS, AND KRICKET KRAP

A colleague of mine was once waiting at a bus stop after attending a meeting of the Canadian Chemical Society. She noticed that a lady standing beside her was suspiciously eyeing the name tag she was still sporting. Finally, unable to control herself, the woman blurted out, "I can't believe that you people are actually advocating a chemical society."

Obviously, to this concerned lady, the word "chemical" was synonymous with evil. "Chemical" meant marijuana or cocaine or heroin. She may even have recalled former us President Ronald Reagan railing against a "chemical society." Our newspapers are also partly to blame for the perceived notoriety of chemicals as they constantly link the word "chemical" with pejorative adjectives. "Dangerous chemical," "poisonous chemical," "carcinogenic chemical," "toxic chemical" are phrases often encountered in the press. It would appear that "useful chemicals," "safe chemicals," and "beneficial chemicals" do not exist. It's time to set the record straight. Let's start by getting a feel for those enigmatic chemicals.

✳   ✳   ✳

The crickets were chirping merrily in their cages on the Georgia farm. Little did they realize that their destiny was to be impaled on fishhooks or ground into pet food. For now, there was plenty to eat; life was good. So they ate and ate and then relieved themselves mightily. "Instead of disposing of the excreta in a landfill, why not sell it as fertilizer?" thought the enterprising farmer. Why not, indeed? So the doo was all neatly packaged and given the clean, scientific-sounding name "cc–84." The problem was that it didn't sell. "Sounds too chemical," people said — it didn't sound like a "natural, organic fertilizer." "Change the name," the inventive farmer thought. "Tell people just how natural the product really is." And so, as "Kricket Krap," the poop of some two billion crickets was soon being disseminated annually through stores and mail-order houses.

✳   ✳   ✳

Robert Tilton was one of America's most popular televangelists (at least until the TV program *Prime Time Live* got through with him). The Reverend Tilton would pray for anyone who sent in a written request accompanied by a donation to one of his charities. The problem was that most of the charities did not really exist. Tilton's flock was actually financing lavish homes, an expensive boat, and even plastic surgery for the preacher. The minister could not deny the evidence, but he did furnish an interesting excuse. He had been reading thousands of prayer requests; his irrational behavior was undoubtedly caused by chemicals in the ink they were printed with. Plastic surgery had also become necessary to correct the damage to his capillaries caused by these chemicals. As far as the $130,000 boat was concerned, it was of course needed to help Tilton relax after the chemical stress he had been under.

* * *

One late-night-TV infomercial was sort of amusing. Its goal was to flog a hair product called "Rio," from Brazil, composed entirely of natural ingredients, that would "relax your curls without chemicals." The host then enlightened us about how "Rio frees you; it doesn't put you in bondage. When you use chemicals you go into bondage." The half-hour sales pitch urged us to be "chemically free" and ended with an enthusiastic testimonial from a formerly frizzy-haired, now neatly coiffed customer who opined, "It's like a death sentence for your hair to use chemicals."

These stories are unrelated, but they do have a common element. Each implies that chemicals are dangerous things and should be either avoided or replaced with "natural" or "organic" substances. Each also makes it clear that there is no appreciation of the fact that chemicals are the building blocks of all matter, that only a vacuum can be "chemically free." There is no recognition here that some natural substances are highly toxic or that the word "organic" is usually bandied about in a meaningless fashion. Above all, there is no recognition of the fact that chemical ingenuity, in less than a century, has transformed a dreary, trouble-filled existence into a brighter, more promising one.

Chemicals are not good or bad. They are just things — the building blocks of the world. It is up to us to decide how we will use them. The same chemical that can kill can also cure. Just consider the following examples. In 1943, German bombers struck a convoy of Allied ships anchored at Bari, Italy. One ship was carrying 100 tons of mustard gas, which spilled into the Bari harbor. Within a month, 83 of the men who had been rescued from the water died. Blood samples from victims were found to contain fewer white blood cells than normal. Since these cells are among those that divide most rapidly, an idea was born. Could mustard gas kill cancer cells? Indeed, the compound is still used in the treatment of Hodgkin's disease.

Botulinum toxin, which can be lethal in microgram quantities, is often cited as an example of one of the most powerful natural toxins known. Yet it has been used in the treatment of cross-eye and in the elimination of facial frown wrinkles. Ammonia is used to make ammonium nitrate, which can be used either as an explosive or a fertilizer. Chlorine can be used as a poison gas, but in its alternate role as a water disinfectant it saves millions from typhoid fever, cholera, and diphtheria every year.

Morphine, a natural product found in the poppy extract known as opium, has ruined countless lives through addiction, but its painkilling effect has also made many disease-plagued lives bearable — the very same chemical used in a different fashion. Today, chemists can actually make synthetic derivatives of morphine that retain the painkilling effects while eliminating the addictive euphoria-inducing properties. The "good" has actually been separated from the "bad." Just like in Robert Louis Stevenson's famous story *The Strange Case of Dr. Jekyll and Mr. Hyde*: Dr. Jekyll uses a chemical to separate the good and the evil sides of his personality. Indeed, the science of chemistry can be termed the "Jekyll and Hyde" science, since it can be used for good or evil. Nitroglycerin used to make bombs can also blast tunnels through mountains or help ailing hearts. Nuclear energy can destroy our world or free us from our reliance on oil. But just as Mr. Hyde's single murder made more news than Dr. Jekyll's entire career of saving children's lives, the negative side of chemicals receives more attention than the positive.

Chemistry, in many people's minds, is linked with the tragedies at Minamata and Bhopal, acid rain, PCBs, dioxins, and toxic waste. There is hardly a thought of aspirin or penicillin or insulin or nylon or lightbulbs or books or television or even underwear — all products of the chemist's ingenuity.

Some of the blame for this does, of course, fall upon the shoulders of the chemical industry, since many highly publicized negative events associated with chemistry can be traced to profit-motivated negligence. But a lack of fundamental scientific education is the real culprit. Children are not exposed to enough chemistry in our elementary schools. It is therefore little wonder that to them chemistry represents the strange and bewildering antics of Beakman on *Beakman's World* or the plots and schemes of the generic "mad scientist" seen in so many cartoons. Chemistry is all bubbling liquids, smoking potions, and — of course — explosions.

Whom do we blame for the stereotypical figure of the mad scientist? This ingrained image may have been inadvertently created by Mary Shelley. Her cleverly crafted novel *Frankenstein* explores the consequences of foolhardy science and leaves us with the message that tampering with nature can lead to unforeseen repercussions. But there is a story behind this story.

"I collected the instruments of life around me, that I might infuse a spark of being into the lifeless thing that lay at my feet." With these words, Victor Frankenstein begins his account of the adventure that would terrify generations of readers. Although Mary Shelley's classic 1818 tale is usually thought of as a horror story, it is actually a thoughtful fantasy about the consequences of science gone astray.

What prompted an 18-year-old girl to write such a dark, scary story about creating life? Works of fiction are often born of real-life experience, so it is interesting to muse about what actual events may have triggered the concept of *Frankenstein*.

First, let's get a couple of things straight. Mary Shelley's Frankenstein is the creator, not the monster. And he is not a doctor. Neither is he a "mad scientist." Victor Frankenstein is a university student who, from a young age, has been obsessed with uncovering the secrets of Heaven and Earth. He voraciously reads the works of the great alchemists, like Albertus Magnus and Paracelsus, who tried to find the secret of everlasting youth. He becomes fascinated with the power of electricity when he sees a tree split by a lightning bolt.

The death of his mother prompts Victor to search even more vigorously for the secret of life. Finally, after numerous failed experiments, he manages to breathe life into the famous creature that he has assembled from body parts. Mary Shelley does not describe the details of that creation; there is no mention of bubbling flasks or electrical generators. All of that was added by the moviemakers. And, quite unlike the Boris Karloff incarnation, Frankenstein's creature learns to think and converse intelligently. It is only when society shuns him because of his appearance that he turns violent. Victor Frankenstein has unintentionally unleashed a scourge on society.

Was Mary Shelley herself worried about what unbridled science might unleash? Perhaps. She had gone to, and been impressed by, a public demonstration of "galvanism." Luigi Galvani had discovered that by touching a severed frog leg with metal instruments he caused the leg to quiver. He interpreted this phenomenon, wrongly, as "animal electricity." Actually, he had accidentally set up a battery with two dissimilar metals acting as terminals and the frog's fluid as electrolyte. This demonstration had quite an impact on Mary Shelley, and she even dreamed of witnessing a stillborn baby brought back to life by electricity.

Mary married Percy Bysshe Shelley, who had left his wife and children for her. They left England because of the ensuing scandal and took a boat tour down the Rhine River, stopping at a castle that had become a tourist attraction based on the exploits of a former inhabitant named Johan Conrad Dipple.

Dipple was a seventeenth-century alchemist who pursued knowledge relentlessly (shades of Victor Frankenstein). Rumor even had it that he had dug up graves and collected cadavers for macabre experiments; he was passionate about finding out how the body worked. He also created "Dipple's Oil," which supposedly prolonged life, and he may have died from tasting his own concoctions: he met his demise foaming at the mouth and convulsing. The name of the castle? Castle Frankenstein.

The Shelleys also stopped at another Rhine tourist attraction — a museum featuring "automata," ingenious clockwork creatures created by master craftsmen. While it is doing them a great disservice to label them elaborate wind-up toys, that is essentially what they were. Some survive to this day and still amaze people with their lifelike antics.

So the stage was set. Mary had been impressed by galvanism. She had visited Castle Frankenstein and learned about Dipple's efforts to create life. The automata she had seen looked alive. It is therefore little wonder that when she, her husband, and two friends, forced inside by the cold Swiss weather, fell into writing horror stories, Mary produced her classic tale of Frankenstein. In so doing, she taught us an important lesson: we must think carefully about the consequences of science, whether we are assembling body parts or molecules.

But she also inadvertently set the stage for the recurring mad-scientist character of books, television, and movies. Victor Frankenstein was not a crazed, goofy scientist, but he was converted into one by the various movie directors who brought his story to the screen. And the image of the mad, self-centered, uncaring meddler surrounded by sparking wires and bubbling flasks has plagued scientists ever since.

Jerry Lewis didn't help matters much with his portrayal of the "nutty professor." The character wasn't malevolent, but it certainly established the enduring stereotype of the dimwitted, bungling chemistry professor. Then there was Fred MacMurray, the "absentminded professor" who invented the fascinating, bouncing "flubber" but couldn't remember how he had done it. Christopher Lloyd's character in *Back to the Future* further crystallized the image of a scientist as a goofy, misguided social outcast.

This image has become so ingrained in our social fabric that movie and television producers feel the need to cater to it whenever a scientist character is called for. Even with the current explosion of children's science programs, there is no getting away from such nerdy, bizarre depictions. The prevailing philosophy seems to be that science cannot stand on its own merit — it has to be sugarcoated, humorized, and musicalized.

Bill Nye, television's "Science Guy," poses for a publicity photo behind a contrived configuration of flasks and beakers that would be alien to any real lab. Meanwhile, Beakman, of *Beakman's World*, rants and raves, displaying his stilted humor in front of incongruously tilted cameras. The pandemonium only subsides during his consultations with Professor I.M. Boring, who is, of course, the stereotypical crazy-looking scientist with a German accent. Is it any wonder that children grow up thinking that scientists are, by and large, eccentrics? Should we be surprised that an Australian survey revealed that 13- and 14-year-olds view scientists as "nerds and losers who devote their lives to hopeless causes and are not accepted by society because they don't want to be?"

The truth of the matter is that science is exciting enough on its own. Imagination, charm, and wit can certainly enhance any presentation, but children do not have to be cajoled into liking science by frizzy-haired scientists or bow-tied nerds. The splendid colors of a rainbow, a bolt of lightning, a rocket blasting skyward, an embryo developing into a baby, a new cancer drug, a biodegradable plastic — all scientific wonders that should stir the imagination. There is no need for giant talking rats or scatterbrained, white-coated hosts with pen-filled pocket protectors to generate enthusiasm.

Straightforward talk about chemicals and their role in our lives can capture students' interest and put adults' concerns into perspective. So, let's give it a shot. You'll never look at that TV commercial about the laxative "that works naturally, not chemically" the same way again. You may also find yourself wanting to argue with Meryl Streep, who, in her role as spokeswoman for the environmental organization the Natural Resources Defense Council, proclaimed, "my grandparents didn't need chemicals to grow food." Streep either comes from a line of magicians or she doesn't realize that all fertilizers are chemicals, be they modern synthetics or old-fashioned Kricket Krap.

AN APPEAL FOR CHEMICAL LITERACY

There were two young men already riding the radio-station elevator when I got in after finishing an on-air stint. "Are you anybody?" one of them blurted out. While I was pondering an appropriate answer to this deeply philosophical question, his crony spilled the beans: "Yeah, he's that guy who talks about chemistry on the radio." This was just the ammunition the philosopher needed. "Oh no, we're locked in an elevator with a scientist," he mocked, before volunteering the information that in high school he had gotten about 2 percent in chemistry, and "that was with cheating."

I've heard this kind of stuff before. After delivering many a public lecture I've been approached by people who somehow feel the need to unburden their souls and tell me, with some sort of perverse pride, that they slept through high-school science classes or that chemistry was the only course they had ever failed. Such comments are emotionally painful to anyone who teaches science. But, worse than that, they imply that poor and unimaginative science teaching may be partly responsible for the frightening degree of scientific ignorance that permeates our society.

Scientific illiteracy is not a laughing matter. Sure, we're amused by silly exam answers suggesting that Benjamin Franklin produced electricity by rubbing two cats together or that one can identify carbon monoxide because it has an "odorless smell." But unfamiliarity with basic scientific principles can give rise to unfounded fears and open the door for charlatans.

Recently, I heard from a gentleman who was concerned that if he slept under an electric blanket he would be "filled with radioactivity," from people who had invested in a Costa Rican company that had discovered a process for turning volcanic beach sand into gold, and from a lady who worried that silicon dioxide in her artificial sweetener would give her breast cancer.

The first two, I hope, require no comment, but the silicon-dioxide question presents an interesting case. Silicon dioxide is just sand. Apparently, the worried lady had confused the word "silicon" with "silicone," which is the name for a type of synthetic rubber that has been used in breast implants. While some problems have been caused by silicone implants, breast cancer has not been one of them. Here a couple of false assumptions led to some very real, but unrealistic, fears.

Why is there silicon dioxide in the artificial-sweetener package in the first place? These sweeteners are so potent that we need to use very little of them. They are mixed with substances such as sand to make packages bigger and handling easier. A little silicon dioxide in our diet is certainly not a problem, but to the uninformed it represents another insult to the body, another "chemical" being foisted upon us.

Oh, yes, those notorious chemicals! Is there any term that is more widely misunderstood? Let me offer some more examples. The Frugal Gourmet, author of bestselling cookbooks, claims that "people don't want to waste time cooking so they go to fast-food restaurants but they lose five years of their lives from eating food with chemicals in it." A chemical-free meal would not be a good deal, unless you like to dine on a vacuum. On a TV talk show, an aromatherapist describes her search for a line of cosmetics that is "relatively free of chemicals." She smells profits; I smell nonsense.

Chemical absurdity has even made it into the courtroom. The prosecutor in a California gang-fight trial described "a situation very much like nitrogen meeting glycerin; it was guaranteed that there would be an explosion of violence." He was probably basing this on some vague notion that nitroglycerin is a potent explosive, but this substance is not made by combining nitrogen with glycerin. Actually, glycerin meets nitrogen all the time quite peacefully: air itself is 80 percent nitrogen.

In a more serious vein, not long ago, cleanup crews dressed in decontamination suits descended on the small American city of Texarkana to deal with a toxic emergency caused by a mercury spill. The culprit was not some careless chemical company — it was chemical ignorance. A couple of teenagers had found a 20-kilogram batch of pure mercury in an abandoned neon-light factory and proceeded to have some great fun with the shimmering substance. They played with it, distributed some of it to friends, spilled it on the floor at home and at school. As a result, eight homes had to be completely emptied of furnishings and six students ended up in hospital, where they had plenty of time to contemplate the dangers of mercury, dangers they should have learned about in high-school chemistry class.

This mercury episode is pretty scary in terms of what is says about science education. But even more chilling is the story of young Nathan Zohner, who won the Greater Idaho Science Fair by getting 43 out of 50 passersby to sign a petition to ban dihydrogen monoxide because it can be fatal if inhaled, it is a major component of acid rain, and it can be found in the tumors of terminal cancer patients. What was this horrible chemical? Water, of course (H_2O).

You've guessed by now that the preceding is an appeal for more and better science education at all levels. We are in trouble when surveyed adolescents reveal that they view scientists as "nerds and losers." We are in trouble when a magazine advises its readers to drink water frequently because "one third of water is oxygen and drinking it will keep you alert." We are in trouble when it is possible to graduate from high school without ever taking a full course in chemistry, physics, or biology.

But there are also some positive signs. High-school science fairs are mushrooming. Some colleges and universities offer programs that emphasize everyday applied science instead of esoteric theory. Perhaps the most encouraging factor is that we educators are blessed with wonderful raw material: many of our students show themselves to be creative, insightful, and perceptive when guided to see science as a fascinating, widely applicable pursuit instead of a compilation of irrelevant and boring concepts and formulas. There is ingenuity out there to be cultured. I recently met a student at a science fair who had developed a way to paint a toilet seat with a luminous chemical so that it could be easily located in the dark. I suspect he won't be signing any petitions to ban dihydrogen monoxide.

THOSE FASCINATING CHEMICALS!

The Lot of Lot's Wife

The tourist guide points to a pillar of stone as the bus leaves the Dead Sea and enters the Negev desert. "That is Lot's wife," he explains in a serious tone. Ears perk up as he quickly relates the familiar Bible story of the righteous Lot and Mrs. Lot, who were warned by the Almighty about the impending destruction of Sodom and Gomorrah. "They would be allowed to leave without fear of being harmed as long as they didn't look back upon the firestorm which would consume the evil cities. But Mrs. Lot's curiosity got the better of her — she sneaked a peek and immediately turned into a pillar of salt. And there she has stood for millennia."

There are chuckles all around, and some of the less sensitive husbands even poke their wives as if to underline the hazards of being too curious. Finally, though, the pillar story is passed off as a bit of tourist-guide fluff. But is that all it is? A chemistry professor at Northwestern University in Chicago has argued otherwise.

In a paper submitted to the prestigious *Journal of the Royal Society of Medicine*, Dr. I.M. Klotz has claimed that there is a scientific explanation for the tale of Lot's wife. In an article filled with equations, formulas, and high-powered technical language, Klotz explains how Mrs. Lot could have literally turned into a pillar of calcite, a form of the common mineral calcium carbonate.

Everyone knows that our bones contain calcium, but fewer are aware that our blood and tissues also contain the mineral. Indeed, our nervous system and heart could not function in the absence of calcium. It is also a well-established phenomenon that when organic matter burns carbon dioxide is produced. Without a doubt, massive amounts of carbon dioxide were released in the inferno of Sodom and Gomorrah.

When Mrs. Lot turned around she got a good whiff of that gas, and this triggered an instant reaction in her tissues, with the calcium forming insoluble calcium carbonate. According to Professor Klotz, she literally turned to stone, dying of "rigor calcium carbonatus."

An interesting thesis. The journal's editors obviously thought so, deeming it worthy of publication. There is only one problem with this fascinating chemical saga: it is utter nonsense. A moment's reflection immediately reveals that the few grams of calcium present in human tissues could never turn a body to stone, even if the reaction with carbon dioxide were a possibility. Could a chemistry professor make such an elementary error? Of course not. Dr. Klotz was not making a mistake — he was making a point.

He wanted to show how easy it is to get nonsense published in the scientific press. The people who review articles submitted to the *Journal of the Royal Society of Medicine* tend to be medical doctors. In all likelihood, they had long forgotten their basic chemistry and assumed that Dr. Klotz's complicated chemical discussion made sense. Klotz undoubtedly enjoyed reading the letters to the editor that focused on the nuances of his theory.

What are we to learn from this bit of mischief? That a degree of skepticism is very healthy when dealing with information. Nonsensical arguments can sound very logical and be very persuasive in the absence of pertinent background knowledge. Perhaps the next time we read about alien abductions, psychic spoon-bending, or the latest dietary supplement that cures all of humanity's ailments, we will benefit from reflecting on the mischievous chemical lot of Lot's wife.

BASKING IN THE LIMELIGHT

Everyone likes to be in the limelight now and then, but who gets the chance anymore? Tungsten light, maybe. Or halogen light. But lime light — no.

Have you ever wondered where the term "limelight" comes from? It has nothing to do with the common green fruit that, coincidentally, happens to be called a lime. It has everything to do with the chemical calcium oxide, which is also known as lime. This white compound, which can be granular or lumpy, has an amazing property. It becomes incandescent when heated.

Before electricity, theatrical stages were bathed in the light generated by heating calcium oxide. A lens fitted in front of the glowing lime focused the light and enabled actors to bask in the glory of the limelight. The light was spectacular, and so was its chemistry. It was also a little scary.

The problem, in the theater, was to find a way to heat the lime to the necessary temperature, and the solution required some very clever chemistry. Starting in the early 1800s, the flame that heated the calcium oxide was produced by burning hydrogen in the presence of oxygen, but this was long before these gases could be purchased in cylinders. They had to be generated on-site.

In those days, the under-stage area was a veritable chemical laboratory. Here hydrogen was made by dropping pieces of zinc into sulfuric acid. The gas was then collected and stored in large, bellows-shaped bags. Oxygen was generated by heating potassium chlorate with manganese dioxide. It, too, was stored in gasbags. The hydrogen and oxygen bags were connected to the limelight by pipes, and when illumination was needed the emerging hydrogen was ignited. Obviously, theater fires were a constant threat.

Today, of course, we don't have to rely on lime for spotlights, but lime itself is still in the spotlight. In fact, it would be hard to picture modern life without it. Lime is made by heating limestone (calcium carbonate) and is widely used in agriculture. It is an alkaline substance, or "base," which can be added to soil to neutralize acidity as well as to increase calcium content. Agricultural liming actually predates the Christian era, and as late as colonial times many farms had kilns in which limestone was converted to lime.

Since lime is the cheapest base available, it has even been used to neutralize acid rain. In Sweden, where the acid-rain problem is particularly serious, the "liming" of lakes is common.

Acid rain is caused mostly by industrial emissions of sulfur dioxide, a gas that can combine with water to form sulfuric acid. The problem is greatly reduced if the sulfur dioxide is destroyed by spraying a lime solution into a chamber through which effluent gases pass before being released into the air.

The largest consumer of lime, however, is the steel industry, which uses it to remove impurities such as silicon dioxide from the metal. Lime has also been a component of cement for thousands of years. The Great Wall of China, the Appian Way, and the Temple of Apollo were all built with lime cement. We still use the substance today.

At water-treatment plants, large amounts of lime are used to reduce water "hardness." Adding lime to the water causes dissolved magnesium bicarbonate and calcium bicarbonate (the "hardness" minerals) to precipitate out as magnesium carbonate and calcium carbonate. The water "softened" by this process will not allow soap scum to form.

Lime even has a nutritional aspect. In Mexico, tortillas are traditionally made by soaking corn kernels in lime water until they become soft enough to be pounded into flour. This not only increases the calcium content of a person's diet, but it also improves the flavor of the tortillas. If the process is changed, the tortillas just don't taste right. Today we know why: they are missing 2-aminoacetophenone, a compound that is formed when lime reacts with the amino acid tryptophan and that is so flavor intensive it is detectable at the unbelievably low concentration of two parts per billion.

In Papua New Guinea, India, and Southeast Asia, lime is used in a more unusual fashion. Here, the chewing of betel nuts is a popular pastime. The nuts contain arecoline, a compound that can produce euphoric effects, and the euphoria is more intense if the nuts are chewed with lime. Apparently, arecoline is more active under alkaline conditions. Unfortunately, the strongly alkaline conditions can also cause oral cancers: oral squamous cell cancer is the most common type of malignancy in Papua New Guinea.

Lime has even been used in glue manufacture. Casein, a protein in whey, reacts with lime to form insoluble calcium caseinate, a substance used to glue wooden airplane parts together during the 1930s. The glue was sold as a white powder consisting of whey, caustic soda (for solubilization), and lime. In a sterile world it would have been a perfect glue, but, like cheese, it softens when degraded by microbes. Soft, like Camembert, the glue ran out of the joints.

Although the applications of calcium oxide are interesting, I've always been amazed by the amount of heat released when lime reacts with water to produce "slaked lime," or calcium hydroxide. This reaction is so exothermic that it can produce temperatures as high as 700 degrees Celsius. For this reason, lime has to be kept completely dry while in storage. If it comes into contact with water, a fire may result — wooden sailing ships occasionally caught fire when water leaked into the hold where lime was stored.

But the most unusual lime story involves kitten urine. A few years ago, fire destroyed a Japanese farmer's shed. It seems he had been storing a bag of lime he intended to use for soil improvement there. Initially, no cause for the fire could be found, but a couple of kittens were discovered next to the bag of lime. They had met an unfortunate end; apparently, they'd answered nature's call in the wrong place.

BOYLE'S LAW AND A HIGH-FLYING ELMO

I had an amusing little encounter with a prospective scientist last time I was flying home from Toronto. Sitting next to me was a small boy who was playing with several bags of peanuts, the bribe he'd extracted from the flight attendant in return for being quiet during the flight.

The boy played happily with the unopened bags throughout the journey, but his face took on a perplexed expression as we landed. The bags had noticeably decreased in volume, prompting the youngster to ask his mother where the peanuts had gone. She had no answer and told her son to stop asking so many silly questions. The lady was obviously unaware of Boyle's Law.

Robert Boyle was born in 1627, in Britain, and was eventually sent off to study at Eton. One evening, while he was outside watching a spectacular display of lightning, he began to wonder why he had not been struck. In a rather unscientific fashion, he concluded that God must have reserved him for some special task. From that moment on, Boyle dedicated himself to demonstrating God's glory by unraveling the secrets of nature.

Boyle became interested in an experiment that had been performed in Germany by Otto von Guericke. In the early part of the seventeenth century, von Guericke had heated a hemispherical copper bowl filled with water until the water boiled. He then fitted a second bowl over the first one, leaving just enough space at the joint to let steam escape. After the heat source was removed, von Guericke discovered that the bowls had become sealed so tightly that two teams of horses couldn't pull them apart. The steam had driven out the air, and when the steam inside the sphere condensed back into a liquid, a partial vacuum was created. The two hemispheres were now held together by the outside air pressure.

All this may sound a little complicated, but the fact is that most of us have carried out a version of this classic experiment in our kitchens. If you remove the lid from a boiling pot and place it on the counter, you'll likely find that it sticks like glue. The trapped steam condenses and creates a vacuum. It isn't surprising that Boyle was fascinated by this effect and was inspired to study the relationship between air and pressure.

Boyle's classic experiment was a marvel of simplicity. He took a J-shaped tube sealed at the short end and proceeded to trap air inside by filling the tube with mercury. He found that the volume of the trapped air varied with the amount of mercury he used and formulated the law that is now studied by every high-school student around the world: the volume of a gas is proportional to the pressure exerted on that gas.

This is exactly what my young traveling companion had experienced. As the airplane gained altitude and the pressure in the cabin decreased, the volume of the peanut package increased. On landing, he could observe the reverse effect.

I didn't feel it was my role to enlighten the young man and his mother about the subtleties of Boyle's law, but this was not the case when I took my daughter to see *Sesame Street Live*. Needless to say, such an outing involved the purchase of a souvenir — in this case, a helium-filled Mylar balloon in the shape of Elmo. Also needless to say, the balloon didn't make it back to the car. Its escape into the great beyond of course elicited tears, but it also prompted a question about what would now happen to Elmo. And this was not such an easy question to answer.

If the balloon had been made of rubber, it would have expanded in size as it floated up in response to the decreasing outside pressure. But temperature decreases with altitude, and gases contract with lower temperatures; this effect may then be expected to shrink the balloon. We therefore have two factors working in opposition. Calculations, however, show that the expansion due to reduced pressure is more significant, and that as the balloon rises it should eventually burst.

This was probably not the fate of the Elmo balloon. Mylar is made of polyester coated with a thin layer of aluminum. It was originally developed to serve as a heat-reflective material in the space program. Mylar does not have elastic properties, but it is extremely strong — so Elmo could rise to great heights without bursting.

In all likelihood, the helium would eventually diffuse through the plastic membrane, and the collapsed balloon would fall back to earth. This, while certainly a comforting thought, did not nix the demand for a replacement Elmo: Elmo number two still exists and is adored, although it is in rather anemic shape due to the loss of helium by diffusion.

Boyle's law has some unusual connections as well. The *New England Journal of Medicine* reports that a lady tourist showed up in the emergency room of a hospital in Frisco, Colorado, complaining of a "swishing" sound in her breasts. X-rays quickly revealed the source of the problem. It seems that the patient had a saline breast implant, which is basically a plastic bag filled with saltwater. Such implants, however, are not completely filled with water and therefore have air pockets. The lady had come to high-altitude Colorado from sea level, and, according to Boyle's law, the air pockets had expanded due to the lower external pressure. The water inside now had room to swish around.

This is a true story, unlike the tale going around about the flight attendant who purchased an inflatable bra and experienced an explosion after takeoff. Although such devices do exist, the small change in volume due to a decrease in cabin pressure is not enough to cause such a spectacular effect. The story is an urban myth that deserves to be deflated.

"Der Schwarzer Berthold"

The inscription on the monument that dominates the town square in Freiburg, Germany, reads simply "Berthold Schwarz." Berthold, the legendary father of gunpowder, is one of my favorite scientists. Constantin Anklintzen assumed the name Berthold when he joined the Franciscan order of monks sometime in the thirteenth century. Because of his interest in black magic, his fellow monks began calling him Black Berthold, or "Der Schwarzer Berthold." He was not, in fact, as much interested in black magic as he was in "black powder," although surely at that time the properties of black powder must have seemed very magical indeed.

Black powder was the earliest form of gunpowder and, according to legend, was introduced into Europe by Schwarz. While this is impossible to confirm — the Franciscan records in Freiburg have long since been destroyed — some historians claim the reason we cannot find any authentication of Berthold's existence is that his name was stricken from all records because he was reputed to have compounded gunpowder with Satan's blessing.

Whether Berthold Schwarz actually lived or not will remain a mystery, but one thing is certain: he did not invent gunpowder. Various formulations based on saltpeter, sulfur, and charcoal were known and used long before the thirteenth century. Credit for the discovery rightly goes to the Chinese alchemists who, three hundred years earlier, published a manuscript describing the flammability of this mixture. They probably made their discovery in the course of their search for elixirs that would guarantee immortality. Taoist philosophy dictated that immortality could be achieved if the opposing forces of yin and yang were brought into perfect harmony within the body. Saltpeter was believed to be rich in yin, and sulfur and charcoal were thought to impart yang properties. Actually, saltpeter (potassium nitrate) is rich in oxygen, which allows the sulfur and charcoal to burn.

The first uses of this fascinating concoction were in Chinese religious ceremonies. Evil spirits, supposedly, did not like the smoke and fire produced by the burning powder. What they probably liked even less was the bang that resulted when worshipers wrapped the powder tightly in paper and ignited it. These early firecrackers were the world's first explosives. The hot gases produced by the combustion process had no place to go and were forced to escape by blowing the confining paper apart, making a loud noise. Red was the usual color of the paper because it was considered the color most feared by evil spirits.

It wasn't long before human ingenuity applied black powder to a more "useful" purpose. By 1044, Chinese warriors were sealing hollow bamboo poles at one end, filling them with the mixture, and igniting them. The burning powder produced hot gases, which escaped from the open end, propelling the bamboo poles in the opposite direction. These primitive rockets must have astounded their enemies.

How news of the discovery of black powder eventually made it to Europe is not clear. What is clear, however, is that the first written account of the phenomenon is attributed not to the legendary Berthold Schwarz but to the very real English monk Roger Bacon. In 1247, Bacon described the explosive nature of a mixture of 40 percent saltpeter, 30 percent charcoal, and 30 percent sulfur. The world didn't find out about this until much later, because the imaginative Bacon, alarmed by the explosive potential of his discovery, enshrouded the formula in a secret code. Some say that it was Berthold Schwarz who eventually solved the puzzle.

Bacon's chemistry was quite poor. His recipe resulted in inadequate combustion, leaving lots of unburned fuel, which was dispersed into the air by the hot gases produced in the form of white smoke. This became a problem as soon as attention turned, as it invariably does, to using the new discovery as a weapon of destruction. By the fourteenth century, black powder was being loaded into iron barrels and exploded to propel iron balls. These first guns were not very efficient because the combustion of the gunpowder was incomplete.

Tinkering with the formula eventually led to the optimal composition of 75 percent saltpeter, 15 percent charcoal, and 10 percent sulfur. Huge cannonballs were manufactured that were capable of knocking down the walls of previously impregnable castles. But soon another problem emerged: supplies of saltpeter were running short.

The only two reliable sources of saltpeter were deposits in India and Spain. These deposits formed over many centuries, saltpeter being one of the end products of the decay of animal and vegetable matter. Then some astute Europeans realized that perhaps it wasn't necessary to look so far afield for supplies of saltpeter. Maybe the barn was far enough.

It turned out that the white encrustations on barn walls that had annoyed many a stable boy were actually saltpeter. The decomposition of manure and other organic wastes had produced the valuable material. Soon, various schemes were worked out to collect saltpeter. Outside villages, "nitre beds" were established where manure and garbage were piled and moistened with urine. Napoleon actually issued an ordinance requiring citizens to urinate on these beds. In Prussia farmers were compelled to maintain piles of organic matter, and in Sweden rural people paid part of their taxes in compost.

With plentiful supplies of saltpeter available, the nature of warfare was changed forever. So were other aspects of life. Coal could be more easily mined and tunnels could be blasted through mountains. And maybe the mysterious Berthold

Schwarz (who couldn't spell his surname quite right) had something to do with it all. At least, I would like to think so.

Movies, Collars, and Billiard Balls

Quick — think of a movie star. Who is it? Elizabeth Taylor? Kevin Costner? Madonna? Miss Piggy? Marilyn Monroe? It's a good bet that the name Fred Ott didn't cross your mind. Yet Ott was the world's first movie star. He didn't make a lot of money, he wasn't besieged by autograph hounds, he didn't make the cover of the tabloids . . . in fact, all he did was sneeze. But in 1889 that sneeze was captured on the world's first film-strip, and the era of the cinema was under way.

Fred Ott was actually a nondescript worker in Thomas Edison's laboratory in New Jersey. He just happened to be in the right place at the right time — the period when Edison and his coworker William Dickson began to tinker with the idea of moving pictures. Still photography, based on the sensitivity of silver compounds to light, was already a well-established process. The concept of moving pictures, however, had had to wait until George Eastman, the founder of Kodak, developed flex-ible-roll film. This breakthrough hinged upon a clever use of one of the most famous substances ever to come out of a chemical laboratory: the world's first plastic, christened "cellu-loid."

Eastman found that by using the solvent amyl acetate, known today as "banana oil" and widely used in the cleaning industry, he could spread the celluloid into a thin layer that when dry formed a flexible sheet. This sheet could be cut, coated with photosensitive compounds, and rolled. Once celluloid film became available, Edison's ingenuity came to the fore. Not only was he a great inventor, but he was also remarkably skilled at improving upon other people's inventions. Unfortunately, the Wizard of Menlo Park was somewhat less skilled at giving credit where credit was due. In any case, Edison developed a camera capable of moving the flexible film past a shutter that opened and closed at the remarkable rate of 48 times a second. Thus Fred Ott's sneeze was recorded for all posterity in stunning detail.

As brilliant as Edison was, he didn't believe in projected movies, an attitude that would eventually cost him dearly, both in terms of prestige and money. He was convinced that flicks would meet their greatest success in coin-operated individual viewers he called "kinetoscopes." He envisaged potential customers forming long lines to view *The Sneeze*, along with other classics such as *Fun in a Chinese Laundry*. The kinetoscope did enjoy some success in peepshow arcades, but it was soon replaced by projected moving pictures developed by the Lumière brothers in France. These, like Ott's sneeze, were also recorded on celluloid film, so it is not surprising that the dawning of movies became known as "the celluloid age" and the first stars as "celluloid personalities."

Indeed, celluloid gave birth to the movie industry, but the birth of the material itself could well form the basis of a feature film. The heroes are a Swiss chemistry professor, Friedrich Schonbein, and an American inventor, John Wesley Hyatt.

In 1846, in Basel, Switzerland, Schonbein carried out an experiment in which cotton was treated with a mixture of nitric and sulfuric acids. There seemed to be nothing remarkable about the washed and dried product; in fact, the cotton looked just like it had before the experiment. But Schonbein was taken aback when he put a lit match to the stuff: it flared up and vanished in a brilliant, smokeless flame, leaving no residue.

The solution to the problem of smoke-filled battlefields, of soldiers unable to see what they were shooting at, sprang to the chemist's mind. Although black powder — the classic mix of sulfur, charcoal, and saltpeter — could certainly propel bullets well enough, its lingering, offensive smoke choked soldiers and gave away their positions. Could "guncotton," as the explosive new substance came to be called, replace black powder? A series of experiments with nitric acid and cellulose, the basic component of cotton, followed. It soon became clear that by varying the amounts of reagents the reaction could be controlled to produce guncotton that would burn at the desired rate. But how could this substance be converted into a powder that could be packed into the barrel of a gun?

Perhaps if the right solvent were found to dissolve the nitrated cellulose, it could be crystallized, like sugar from water. But the nitrated cellulose did not crystallize. When the solvent in which it could easily be dissolved, a mixture of ether and alcohol, evaporated, a clear, plasticky film was left behind. It took some 40 years to develop the technology to convert this residue into gunpowder, but the film left behind after the evaporation process was quickly put to use.

A Boston medical student named J. Parker Maynard suggested that this "sticking plaster" be used as a waterproof coating on cuts and surgical wounds. A solution of nitrated cellulose in ether alcohol, which came to be known as "collodion," did find widespread use as the first Band-Aid. Just imagine people's amazement when they applied the viscous solution to their cuts and watched it turn into a clear, waterproof dressing. But it was a bottle of collodion accidentally spilled by the American inventor John Wesley Hyatt that provided the right chemistry for the development of celluloid, the world's first plastic.

Have you ever wondered what people did with their leisure time before VCRs, PCs, and Nintendo? They read, they talked, they played games. In the nineteenth century, billiards was the rage, but a problem cropped up as the game's popularity grew.

It had to do with, of all things, a shortage of elephants. The ivory from elephant tusks was the only material suitable for billiard balls, and there wasn't enough of it to go around, so the New York firm of Phelan and Collander offered a prize of ten thousand dollars, a great deal of money at the time, to anyone who could come up with an artificial substitute for ivory.

Word of this substantial prize caught the attention of a young Albany, New York, inventor who had already tried to make checkers and dominoes out of pressed wood pulp and shellac. Then one day, or so the story goes, he cut his finger and immediately went to find some collodion. When he opened the cupboard where it was kept, he saw that the collodion had spilled. As he attempted to clean up the mess, he realized that the liquid had hardened into a solid mass. Could this stuff be molded into the shape of a billiard ball? It turned out that it could, but it took Hyatt seven years to perfect the technique. The secret lay in combining collodion with camphor — a substance that had been isolated from a Taiwanese tree — before molding with heat and pressure. The camphor acted as an internal lubricant (or "plasticizer," as we now call it), allowing the long molecules of cellulose nitrate to slip by each other during the molding process. Hyatt named the new material "celluloid."

Unfortunately, celluloid did not have the same elasticity as ivory and therefore never really panned out as a substitute for billiard balls. Still, as the first moldable plastic it found a variety of other uses. Hyatt and his brother set up the Albany Dental Plate Company and revolutionized the manufacture of dentures. Their product was great for everyone except tea-drinkers: celluloid softened in hot liquid, literally curling the teeth. Celluloid did, however, replace horn, hoof, ivory, tortoiseshell, and hard rubber in brushes, knife handles, jewelry, and piano keys.

Detachable celluloid collars and cuffs became popular. Since gentlemen's shirts were hidden by vests, they could wear the same shirt for extended periods, simply removing and rinsing the cuffs and collars when they got dirty. One of those new-fangled collars could be washed many times and still hold its shape — as long as its wearer didn't get too hot under the collar, that is. Celluloid was highly flammable; in fact, in the space of 36 years, the main celluloid factory in Newark, New Jersey, was the scene of some 39 fires and explosions, resulting in at least nine deaths and numerous injuries.

Due to its flammability and the invention of better plastics, celluloid soon disappeared from the market. This first "miracle plastic," for which people once held such high hopes as a possible substitute for ivory in billiard balls, now has only two commercial uses. Apparently, it is just the right stuff for guitar picks, and no other material can duplicate the resilience of celluloid in Ping-Pong balls. A failure in billiards but a success in Ping-Pong — well, that's the way the celluloid ball bounces.

JEANETICS

The video camera in the Spokane, Washington, bank caught the robbery on tape. While the robber was wearing a mask and could not be identified, his jeans were clearly visible. So FBI scientists quickly went to work to see if the criminal could be identified by his pants.

The seams in a pair of jeans, they had discovered, always show a characteristic pattern of light and dark lines, a consequence of how the operator pushes the fabric through the sewing machine. In some places the fabric bunches up and in others it stretches, resulting in the cotton wearing out at different rates.

A suspect was eventually arrested, and his jeans were sent to an FBI lab. There, forensic scientists found over 20 patterns of fading on the seams that matched those seen on the jeans in the videotape. The district attorney thought this was enough evidence for a conviction.

The defense, of course, didn't agree. Their expert witness, a jeans exporter, testified that such patterns were common to all jeans and did not have unique features. He brought 34 pairs of jeans to court in order to prove his point, and, indeed, casual visual inspection showed that they had virtually identical seams. The defense's confidence was soon shattered, however, when FBI experts showed that they could readily distinguish the suspect's jeans from any of the exporter's 34 pairs. The defendant was convicted.

And this was not the only time that denim had played a part in a criminal conviction. When New York police investigated a robbery/murder at a video store, they found a bullet imbedded in the ceiling in addition to the bullet that had killed the store's clerk. This bullet had human blood and tissue on it that did not match the victim's, along with fiber fragments from a material that appeared to be denim.

A suspect was soon found. He looked ill and would not answer questions. The police, deciding to search his apartment for any denim clothing that might match the fiber found on the bullet, found a pair of bloody jeans with a hole in the seat. Analysis of the blood on the jeans revealed the presence of an enzyme, phosphoglucomutase, that matched exactly a sample of this enzyme found on the bullet. Phosphoglucomutase is a protein that can be used as a genetic marker since its molecular structure differs slightly from person to person.

The suspect also had a burn mark on his leg, and this was tested by means of a reagent that changes color in the presence of chemicals called nitrites. All types of gunpowder contain nitrites. In fact, it is the incorporation of nitrite groups into certain molecules that renders them explosive. The test for nitrites turned out to be positive, and it helped police to nail down the exact sequence of events. The robber had struggled with the clerk and had dropped his gun, which then hit the floor and went off just beside his right leg. The bullet went straight up, ripped through the robber's rear end and lodged in the ceiling. When confronted with the evidence, he confessed and was sentenced to life imprisonment. One suspects that for a while, at least, he didn't exactly "sit" in prison.

It isn't only criminals who have gotten into trouble because of their jeans. In the 1980s, there was a mysterious increase in occurrences of a condition called "meralgia paresthetica," characterized by pain in the thigh. Tight jeans had come into fashion, and some people tried to achieve that "painted-on" look by wearing their jeans while sitting in a full bathtub — their aim was to shrink the garment to the contours of the body. This process worked, but it turned out to be a real pain in the ... thigh.

Believe it or not, this shrinking technique was not pioneered by fashion-conscious teenagers in the 1980s, but by gold miners in the 1850s. The California Gold Rush of 1849 attracted many fortune seekers. Among them was young Levi Strauss, who had arrived from Bavaria with visions of staking a lucrative claim. To raise the necessary money, he had brought along some bolts of canvas, which he intended to sell to miners as tent material. When one of his customers complained that what the miners needed were strong pants, not strong tents, Strauss swung into action.

He fashioned a pair of crude canvas overalls for the gentleman, and the rest, as they say, is history. Other miners began to seek out those wonderful Levi Strauss pants, and a trademark and an industry were born. At first, for 22 cents a pair, you got "one size fits all"; you'd then have to sit in a watering trough until your pants fit.

As the business began to boom, Strauss switched to a more durable material, a type of cotton he imported from the town of Nîmes in France. Since it came *from* ("de" in French) Nîmes, it became known as "de-nim." When his customers complained that the beige-colored material showed stains too readily, Strauss dyed his denim with indigo, a common plant-derived blue pigment.

Another criticism — that the pockets ripped when they were stuffed with gold ore — was addressed by Jacob Davis, a tailor. He came up with the idea of reinforcing the pockets, as well as the crotch, with copper rivets. Davis informed Strauss of his discovery, and the two ended up taking out a patent on the rivet idea. The crotch rivet was eliminated when cowboys, who normally wore no underwear, found that crouching around a campfire exposed the rivet to the full heat of the flames and caused burns in a rather sensitive area of their anatomy.

Jeans, of course, are no longer just a rugged work garment — they have become fashion wear. These days, the older they look, the more appeal they have. For customers who are too impatient to break in their jeans, scientific ingenuity has given rise to stone-washing, a process that can duplicate years of wear. Manufacturers have actually placed the jean fabric into large industrial washing machines with volcanic rocks. The rocks rub against the denim's surface, abrading some of the cotton yarn. Since the indigo used to color the fabric does not penetrate the surface, the undyed white cotton underneath begins to show through.

A clever American inventor actually sells "the authentic stone" for home production of stone-washed jeans. All one has to do is rub a freshly washed pair of jeans with the stone to produce the "authentic effect." (A note to teenagers: your parents will not appreciate it if you attempt to do this in the washing machine.)

Then there are "acid-washed jeans," which are really misnamed. The washing is actually done with porous volcanic rocks that have been treated to absorb bleach. When added to the industrial jean-washing machine, they contact the material in a random fashion, destroying the indigo wherever they happen to touch. The actual type of rock used is a proprietary secret. The pockets of all these jeans are checked before distribution to ensure that rival manufacturers do not find an errant stone.

And what's on the jean horizon? Certain manufacturers are using an enzyme called "cellulase," which breaks down some of the cellulose in the jeans to produce premature fading. Genetic engineers are working on eliminating the need to dye denim: they are attempting to insert into the genome of cotton plants the genes that are responsible for producing the blue color in indigo plants. We may soon be wearing jeans with real blue genes.

JOHN DILLINGER, FAKE SLUSH, AND COUNTERFEIT MONEY

I'm a carbohydrate chemist by training, so I have always had a certain fondness for starch. One of the first things I ever learned about this remarkable substance is that it turns blue when it comes into contact with iodine. I remember treating various foods — ranging from pasta to potatoes — with an iodine solution to test for the presence of starch and observing the telltale blue color.

Chemically, the formation of this blue color is fascinating.

All plant starches are giant molecules formed by linking from 60 to 6,000 glucose units. These glucose "monomers" can actually be joined together in two distinct ways: if they are lined up in a straight chain, we have "amylose starch," which is water soluble; if they are linked in a branched fashion, the resulting starch is insoluble in water and is known as "amylopectin."

Only amylose, which makes up roughly 20 percent of starches, produces a blue color when iodine is applied. The long amylose chain is coiled in the form of a helix, and iodine molecules just happen to be the right size to fit into the central open space in the helix. The resulting complex of amylose and iodine is blue. Neat stuff, but it always seemed to me that when I was describing this aspect of chemistry to students or the public, most of my audience did not share my excitement about coiled molecules. I guess it didn't seem particularly relevant. So you can understand my delight some years ago when I received a new textbook to review that featured a little historical footnote about starch, iodine, and John Dillinger.

John Dillinger was the notorious American gangster and bank robber who, during an 18-month period in 1933–34, got away with over a million dollars in cash. As a result, he was declared public enemy number one by FBI director J. Edgar Hoover. The chemistry text recounted how Dillinger, while serving a jail sentence, had asked for a potato, which the guards had provided; he deliberately cut his finger and asked for some iodine to disinfect the wound. The gangster then carved the potato into the shape of a little gun and painted it with iodine until it resembled blue steel. Pointing this potato gun at a guard, Dillinger made his famous "chemical escape."

"What a great story," I thought. This was the hook that I could use to capture people's imaginations, leading them to an understanding of amylose molecules and a recognition of the importance of scientific education: "The next time you're in jail, just get a potato, some iodine and . . . ." I even devised a little demo to go along with the story — a colorless solution turned yellow, like a potato, then blue to represent the reaction with iodine. This little skit has been most successful in stirring up some interest about the chemistry of starch.

Not long ago, the Arts and Entertainment network advertised an upcoming feature on John Dillinger to be aired on its wonderful *Biography* series. This I had to see. Maybe the show would reveal more details about the chemical escape, maybe even a glimpse of the potato gun. Alas, it was not to be: I was to learn from the documentary that there had been no "chemical escape." Dillinger had, in fact, escaped from the high-security Crown Point Jail in Indiana. He later recounted how he had fooled the guards with a fake gun he had fashioned out of wood and colored with boot polish, but apparently this aspect of the story is suspect. It seems that someone had actually smuggled a real gun into the prison, but Dillinger had concocted the fake-gun tale to ridicule the guards and boost his status as a living legend.

So John Dillinger was not versed in the chemistry of starch after all, but he did know something about acids: at one point, he forced a plastic surgeon to treat his fingertips with concentrated acid to remove his fingerprints. It may have seemed like a good idea, but it didn't work. The skin grows back, and so do the fingerprints. Indeed, after Dillinger was gunned down with very real FBI weapons, his body was identified by its fingerprints.

While the Dillinger story turned out to be false, it doesn't mean the starch-iodine reaction has no practical relevance: it has been used to catch counterfeit-slush dealers. "What," you may ask, "is a counterfeit slush dealer?" Slush is a flavored ice concoction designed to satisfy the cravings of children with uncritical tastebuds. It is dispensed from a machine that is usually rented from a distributor with the proviso that the slush mixture must also be purchased from the same source. Bootleg slush manufacturers pay no heed, but now at least one distributor has come up with a clever system to keep his clients in line. A little starch is added to the mix and can be detected by the addition of iodine. The distributor's serviceperson, who appears regularly to maintain the equipment, just adds a few drops of iodine, and if no color is produced, the equipment is removed — the slush seller may even be sued for breach of contract.

Well, maybe counterfeit slush is no big deal, but counterfeit money is. Here, too, at least as far as US currency is concerned, the starch-iodine reaction may be employed to catch criminals. The manufacture of the paper currency is printed on is a highly secretive business; there are all kinds of built-in safeguards against counterfeiting, including the removal of all traces of starch from the paper. Apparently, counterfeiters have not managed to copy this proprietary process, and so law enforcers can detect fake currency by drawing on it with a special pen that dispenses iodine. The appearance of a blue color signals forgery.

And there is one more interesting application of the starch-iodine reaction: it is used in the treatment of iodine poisoning. Since starch binds iodine so well, the accepted procedure is to administer starch orally. In this case, the blue product will make the patient feel less blue.

PLAYING WITH CHEMICALS

If you're looking for a gift for a youngster, may I suggest a chemistry set? I got my first one when I was about 12, and I can still recall the thrill of changing water into "wine" for the first time. Since then, I have performed hundreds of chemistry "magic shows" featuring many of the basic reactions I learned in my younger days, and I sense that today's audiences still find them just as enchanting as their predecessors.

It is certainly appropriate to begin any discussion of chemistry sets with a tip of the hat to an Englishman named Michael Faraday, one of the greatest scientists of the nineteenth century. Faraday discovered benzene as well as the process of electrolysis, and he made the first electrical dynamo. He gave wonderful public lectures on chemistry at the Royal Institution in London, and his Christmas lectures for children are still regarded as classics.

Many a child was turned on to chemistry by watching Faraday's performances and responded especially to his encouragement to "work at home." Chemistry sets, then known as "chemical amusement chests," became so popular that shops catering to the trend popped up all over London. In fact, so much interest was generated in this fashion that when 180 noted scientists were asked, in 1874, what had sparked their interest in science, many claimed it was the "chemical amusement chest" of their youth.

Today, sadly, children don't seem to possess such curiosity and enthusiasm. Movies, television, video games, computers, and a mind-boggling array of toys all vie for their attention. Kids who have grown up with lasers, special effects, and Nintendo are less likely to be amused by chemical color changes than were their Victorian counterparts. The association of the word "chemical" with toxicity has also made parents think twice about giving their children chemistry sets. This is unfortunate, because a knowledge of chemistry fostered by experimentation is more vital and more useful now than ever before.

There are all kinds of interesting and stimulating experiments for children to carry out. If instructions are carefully followed and common sense is used, chemistry sets can safely provide hours of fun-filled learning. Common sense, however, is not as common as we might like. A few examples of the improper use of chemistry sets highlight this point.

One of the classic children's experiments is the making of a chemical weather predictor. This project hinges on the fact that cobalt chloride is blue when it is dry and pink when it absorbs moisture — such a substance is said to be "hygroscopic." Salt, for example, also has this property, and so we try to counter the stickiness caused by the absorption of moisture from the air by putting a few grains of rice into the salt shaker. A chemical weather predictor is made by wetting a piece of filter paper with a solution of cobalt chloride and allowing it to dry. As the paper dries, its pink color changes to blue. If the paper is then placed by a window or on an outside wall, the color will change back to pink if the air becomes humid. Many novelty items, as well as commercial products like gasoline line filters, are based upon this phenomenon.

But children can also get other ideas — such as putting cobalt chloride into your sister's juice to see if it changes color. When a mother in England asked her six-year-old boy to prepare a black-currant drink for his four-year-old sister, he recognized a perfect opportunity for experimentation and promptly added about 2.5 grams of cobalt chloride to the beverage. His mother, however, suspected that he had added some soap. She told him to drink the glass of juice himself, which the boy promptly did before heading off to school. All morning he complained of abdominal pain, and when he began to vomit he was promptly taken to the hospital. He was given an emetic to clear his stomach and released without any further problems after 48 hours. The level of cobalt in his blood was extremely high, but it returned to normal without further treatment after a couple of months.

Others have not fared so well. A 14-year-old girl had to be given a drug (EDTA) to remove cobalt from her body after ingesting (for unknown reasons) a massive amount of a cobalt compound from her chemistry set, and a 19-month-old boy died after drinking a cobalt chloride solution. This is not surprising in light of what we know about the toxicity of cobalt compounds. When cobaltous sulfate was added to beer in the Province of Quebec in the 1960s as an antifoaming agent, it was quickly found to be responsible for at least 20 deaths due to heart damage.

Care must be taken when using chemicals. A final story illustrates the point. An 11-year-old girl became intrigued with growing crystals, and she prepared a solution of copper sulfate according to the instructions given in her crystal-growing kit. Before going to sleep, she placed a glass of Ribena (the same black-currant drink the English boy concocted for his sister) on her night table. Waking up thirsty in the night, she gulped down the wrong solution. It was a fatal mistake.

Such isolated episodes, coupled with the fear of lawsuits, have resulted in perhaps less interesting but virtually foolproof chemistry sets. I know this, because I recently searched for a set I could use to introduce my four-year-old daughter to the delights of chemistry (you can never start too early). I was unprepared for what I found. My own first set was called "Exploring Chemistry"; now the salesclerk led me to a shelf bearing kits named "Stinky Smelly Hold-Your-Nose Science" and "Icky Sticky Foamy Slimey Ooey Gooey Chemistry." Times have changed. It seems that in order to capture the interest of children these days, manufacturers have to promise "the grossest, slimiest, foamiest, weirdest things you've ever seen."

Anyway, I bought the "Icky Sticky" set. At home, we undertook to follow one of the procedures, mixing laundry starch, food dye, and white glue to make "Greasy, Grimy

Gopher Guts." The resulting chemistry was actually quite interesting and involved the formation of giant molecules, or polymers. I recognized the experiment because it had also been featured in my early chemistry set — under the title "Building Interesting Molecules."

In fact, it was under this name that I presented the experiment to my daughter. She was thrilled until I mentioned that the instruction manual had referred to the product as "gopher guts." "Yuuuuckk! Who wants to make that!" was her rapid reply. And then it was back to Barbie and videos. I went out to look for a more traditional chemistry set that featured changing water into wine rather than "grossing out your friends with fake blood." After all, it's the same chemistry.

RADAR, HULA HOOPS, AND PLAYFUL PIGS

It was an accidental discovery, but it changed the eating habits of the nation, altered the outcome of the Second World War, and produced hula hoops, toys for pigs, and burpless babies. And let's not forget Frisbees and Barbie dolls. We're talking about polyethylene. The chemical name may not ring a bell, but almost everyone is familiar with this material in the form of shopping bags, squeeze bottles, margarine tubs, cling film, or the tags on pillows and mattresses that bear an ominous warning about the legal consequences of their removal.

Our story begins one Monday morning way back in 1933. Two organic chemists working at Imperial Chemical Industries (ICI) in England began the week by checking up on an experiment they had begun the previous Friday. Their research focused on chemical reactions at high pressures, and they had designed an experiment in which a petroleum-derived gas called ethylene was mixed with another reagent in a pressurized

cylinder. Much to their surprise, the tank gauge showed no pressure on Monday morning. They feared their reagents had leaked out, but closer examination revealed the presence of a white powder in the reaction vessel. The small ethylene molecules had joined together to form giant molecules of poly-ethylene. The two chemists had discovered a new plastic.

Within a short time, the techniques needed for mass production had been worked out and the only thing left to do was to determine a use for the new material. At this point, the British Telegraph Construction and Maintenance Company found out about polyethylene and decided to try it as an insulator for underwater cables. By 1938, a telephone cable had been successfully laid between the British mainland and the Isle of Wight.

And then war broke out. The Allies had been secretly working on radar, but they had failed in their attempts to install the equipment in airplanes. The device required a great deal of specialized insulation, but the materials available at the time were all too heavy for use in the air. Polyethylene was light and fit the bill. Soon, the Royal Air Force was flying missions with the help of radar, and British pilots sank over a hundred German submarines (U-boats) within a few weeks. Hitler ascribed the "temporary" setback to that "single technical device." The Germans worked feverishly to develop airborne radar equipment of their own but could not do so without the polyethylene technology. The tide of battle had been turned. After this, war uses for polyethylene mushroomed.

Earl Tupper learned about the material while working as an engineer at the Dupont Company. He hit upon an idea that would permanently change people's eating habits by making leftovers easily storable. His brainchild, of course, was Tupperware, a line of molded polyethylene containers that were flexible, strong, and capable of providing an airtight seal.

Research soon yielded many new varieties of polyethylene. A special catalyst developed by two eventual Nobel Prize winners, Karl Ziegler and Guilio Natta, resulted in high-density polyethylene, which was stiffer and stronger than the original substance. Large-scale manufacture, though, led to problems, as the plastic cracked easily. Luckily, there was a commercial use for poor-quality polyethylene: the hula hoop. The hoop took America by storm. Rock and roll was transforming the nation, and everyone wanted to swivel their hips like Elvis. The hula hoop was the perfect teaching device. By 1958, twenty thousand hoops were rolling off assembly lines every day. When someone established a world record by twirling fourteen hoops at the same time, it received extensive publicity. But for some, this was just too much hoopla. Many fundamentalists opposed the hoop because of the sexual innuendo of the gyrations. Indonesia went as far as banning the twirling, fearful that the motion might stimulate inappropriate passions.

Eventually, the problems with the large-scale manufacture of the material were worked out, and we now have a variety of polyethylenes for various purposes. Garbage bags, shopping bags, and cling film can all be made of polyethylene. The era of the burpless baby was born with the introduction of the small collapsible bags used as bottle inserts. Babies would no longer be sucking in air while being fed.

Even pig farmers have benefited from advances in polyethylene technology. When pigs are raised in the close confinement of piggeries, they tend to annoy each other. They nibble on their confreres' tails. This can lead to infections, so pig breeders often put rubber tires and bowling balls in pigpens to distract the creatures from their peers. Furthermore, when the pigs start pushing these toys around they get exercise, which promotes weight gain and prevents the pork from becoming watery due to improperly developed muscles. Now there is a new, improved pig toy, thanks to high-density polyethylene, which can be molded into balls that may be inflated with water. Contrary to what we may think, pigs don't actually like dirt; these balls are easily washed, since there are no holes or threads to collect dirt. The pig balls can even be adjusted in size as the pigs grow.

So polyethylene helped the Allies win the Second World War, and it gave us shopping bags, Tupperware, and toys for pigs. One of the substance's more recently developed uses has been in the manufacture of artificial hip joints, which are made of ultra-high-molecular-weight polyethylene. These joints may turn out to be just the thing for those who suffer from polyethelenitis due to overzealous hula-hoop twirling.

And just when you think you've covered the spectrum of weird and interesting uses for polyethylene, another one comes along. I like to spice up my chemistry presentations occasionally with a little magic and a little humor. So one day I had an idea. Real magicians produce doves. Wouldn't it be appropriate for a "chemical magician" to produce a plastic dove instead? After all, plastics represent some of the most important chemicals we have. Why not begin a lecture on plastics by magically producing a "synthetic" dove? It took me a while, but I finally found an appropriate creature. It could even flap its elastic-powered wings and fly. Guess what the wings were made of? That's right, polyethylene! Now if we could make these wings bigger, they could perhaps be attached to pigs. I bet the animals would enjoy that even more than rolling around those polyethylene pig balls. And when will we finally run out of interesting uses of polyethylene to talk about? The answer is obvious. When pigs fly.

THE GREAT PHENOL PLOT

"Kills Germs by Millions on Contact." Most of us recognize this as the famous Listerine mouthwash and gargle slogan. Joseph Lawrence, an American physician, developed the familiar yellow liquid in the late 1800s and named it after the brilliant British surgeon Joseph Lister. No, Lister did not have bad breath. The product was named in his honor because Lister is widely considered to be the father of antisepsis, the science of preventing infection.

Lister knew that fractures that broke through the skin would often become infected, whereas those that did not pierce the skin healed nicely. The prevailing opinion at the time was that the exposed tissues were affected by oxygen in the air; oxygen would break down the components of organic matter in a wound and generate pus. The common method of excluding oxygen in Lister's time was to dress wounds with tight bandages. These dressings actually encouraged bacterial growth and resulted in a virtually indescribable stench on hospital wards. Many doctors believed that the odor caused the infection and that it was directly responsible for the extremely high death rate following surgery. Yet, incongruously, nobody tried to solve the problem by eliminating the smell. The sole source of light in this darkness was Florence Nightingale, the legendary Lady with the Lamp, who espoused a doctrine of soap, warm water, and sunshine but was largely ignored.

Then came a breakthrough. A professor of chemistry, Thomas Anderson, introduced Lister to the ideas of Louis Pasteur, who had shown that rotting and fermentation could occur in the absence of oxygen, as long as microorganisms were present. Furthermore, the microorganisms could be killed by heat. This really struck a chord with Lister, who had never believed in the oxygen theory anyway. Indeed, he had fantasized about some sort of invisible dust settling into wounds. Lister immediately designed an experiment. He took some fresh urine, heated it, and sealed half of it in a glass tube, leaving the other half exposed to the air. When he smelled the samples in the morning, the one that had been exposed to the air reeked while the sealed sample was odorless. Evidently, microorganisms from the air had infected the open sample.

Since heating a patient was of course not a viable approach, he wondered if the germs could be killed with appropriate chemicals. Lister thought of carbolic acid, or phenol, because he knew that it had been used to clean foul-smelling sewers. He also knew that when the treated sewage was used as fertilizer, the cows grazing in the pastures it had been spread upon did not become infested with parasites, an otherwise common occurrence. Perhaps the stuff that destroyed the smell and the parasites could also kill Pasteur's microorganisms.

Lister got some carbolic acid from Anderson and tried it on a boy who had been run over by a cart and had suffered an exposed fracture of the tibia. The child recovered with no complications. Soon Lister was washing his instruments with phenol, and he also developed a sprayer with which he could mist disinfectant throughout the operating room. The results were immediate: the mortality rate from amputations dropped from 50 to 15 percent. Nevertheless, Lister had to deal with a great deal of skepticism, because the germs, or "little beasts" as some called the microbes, were not readily observable. But, in 1867, the prestigious British journal the *Lancet* accepted Lister's article on the prevention of infections, and the era of antisepsis was launched. Phenol would save thousands of lives — but it would also end many, for scientists quickly discovered that phenol could be converted into the potent explosive trinitrophenol.

Every student of history knows that the First World War began with the shooting of Archduke Ferdinand in Sarajevo, but what many don't realize is that the United States did not enter the fray for another two years. During this time, the Germans made various attempts to keep the United States out of the conflict and to prevent it from offering technical help to Germany's enemies. One of the most intriguing schemes hatched by the Germans has come to be known as the Great Phenol Plot.

The "plot" revolved around an attempt to corner the market for phenol and the powerful explosives that could be derived from it. The phenol manufacturing industry was, at that time, centered in England, and after the outbreak of the war, most of the available phenol was channeled into munitions production. This resulted in a decline of phenol exports and caused a phenol shortage in the United States. The Germans were worried that the shortage would spur the Americans to start manufacturing their own phenol. American efficiency would likely yield massive amounts of phenol, some of which would undoubtedly find its way into the hands of Germany's adversaries. The German ambassador to the United States was therefore assigned the task of preventing American chemical companies from supplying phenol to the Allies already engaged in fighting Germany.

For this unique undertaking, the German ambassador enlisted the help of Hugo Schweitzer, a German chemist living in

New York. At first, Schweitzer's job was not difficult because the American phenol industry was hardly significant. But then, in 1915, he ran into an unforeseen but monumental problem: the American genius Thomas Edison. Edison, by this time, was heavily into the business of marketing his phonographs and records. The records were made of a plastic called Bakelite, which was made from phenol. Since the United States had no access to the British supply of phenol, Edison decided to make his own, and the brilliant inventor devised such an efficient manufacturing process that he had more than enough phenol for his phonograph records — in fact, he had begun looking for a market for his excess.

Schweitzer, charged with keeping Edison's excess phenol from being turned into explosives and shipped to Europe, racked his brains. Thanks to his background in chemistry, he soon came up with a solution. Phenol, he knew, was used in the making of aspirin. Furthermore, the Bayer Company, operating in the United States, was already feeling the phenol pinch. Maybe he could convince Edison that the most humane thing to do with his surplus phenol was to sell it to the aspirin maker. Who could object to that?

Not Edison, apparently: he signed a contract with Schweitzer that enabled the spy to divert a supply of phenol equivalent to four and a half million pounds of explosives. But then the US Secret Service caught wind of the plot. Schweitzer could not be jailed because the United States was not yet at war and Germans were still allowed to buy any American product they desired, but when Edison learned the real motive behind the Bayer phenol purchase, he decided to sell his excess phenol to the US military. The Great Phenol Plot was spoiled. Still, we must grudgingly recognize the ingenuity of the German spy, Schweitzer, whose knowledge of the chemical link between aspirin and phenol almost caused the Allies one giant headache.

The Rise and Fall of a Gas

The Perkin Medal is one of the most prestigious prizes awarded in the field of chemistry. It is presented annually at a formal gala celebration, which is capped off by the winner's address. Most award winners deliver the standard speech — they thank everyone in sight and reminisce about their long careers in chemistry — but 1937 winner Thomas Midgley's presentation was different. Midgley began his address by inhaling some Freon gas, which he then exhaled through a tube, extinguishing a lighted candle. It was a sensational demonstration of the nontoxicity and nonflammability of the gas, but why did Midgley indulge in such a theatrical performance at a black-tie academic function? To convince the chemical community that Freon, or dichlorodifluoromethane, was an ideal substance to use as a refrigerant.

Midgley was actually being honored for his discovery of the antiknock properties of tetraethyl lead in gasoline, but his pet project at that point was the replacement of problematic ammonia and sulfur dioxide in refrigerators. The inventor was having trouble convincing manufacturers of the safety of Freon and hoped that his demonstration at the highly publicized awards dinner would help muster support. The ploy worked, and soon refrigerators and air conditioners were humming away, loaded with Freon instead of toxic ammonia and sulfur dioxide. Consumers no longer had to worry about pipes corroding and dangerous gases escaping. Refrigerator sales went up and food poisoning went down; everything seemed right with the world.

But then the sky started to fall — or, at least, it opened up and allowed damaging ultraviolet light to pass through. By the 1970s, concerns were being raised that the chlorofluorocarbons — or CFCs, as they had come to be known — were not so benign after all. Escaping from spray cans, refrigerators, and air conditioners, they migrated to the upper atmosphere where they went to work destroying the ozone layer that protects us from excess ultraviolet light. Soon, Freon-propelled spray cans were banned, and plans were drawn up for the eventual phase-out of all CFCs: the hero was turning into the villain.

Thomas Midgley did not live long enough to witness the fallout from his invention — a shame, for his brilliant mind would have geared up to seek a solution. The noted chemist fell victim to polio and was confined to bed. Still mentally alert, he devised a pulley system to get out of bed, but one day he accidentally got entangled in the ropes and strangled him-self. In my opinion, science lost a champion that day, but not everyone would agree with me. A few years ago, I had the dubious pleasure of attending a supposedly educational play sponsored by La Ligue des Femmes du Québec, which por-trayed Midgley as a fiend who got just punishment for a life-time of polluting. Filled with memorable lines — like, "Thomas is dead and buried. He has stopped polluting" — the skit ended with the admonition that we must guard against being as stupid as Thomas Midgley. It seems the educators responsible for this presentation need some educating.

In the context of the 1930s, Midgley's contributions were spectacular. No one could have predicted that 50 years later those trailblazing CFCs would blaze a hole through the ozone layer. At the time, lack of refrigeration and consequent food poisoning was a major problem. Midgley's contributions to the science of refrigeration undoubtedly saved many lives. His portrayal as an uncaring rogue just demonstrates the ignorance of all those involved in this absurd, antiscience play.

There are, in fact, real scoundrels in the CFC saga. The con-trols on the production and use of CFCs imposed by the Montreal Protocol of 1987 have given rise to a profitable new business: the large-scale smuggling of chlorofluorocarbons. These substances are in demand because the legal alternatives that have been developed, the so-called hydrofluorocarbons (HFCs), require extensive modification of existing refrigerators and air conditioners. It costs anywhere from three hundred to eight hundred dollars to refit a car's air-conditioning system to accommodate the more environmentally friendly HFCs.

Malfunctioning air conditioners often lose Freon, and it is obviously far cheaper to repair faulty systems and refill them with Freon than to modify them for HFC use. Currently, recycled Freon can still be used in North America, and so can CFCs that have been stockpiled. Further manufacture of these chemicals is illegal, so supplies are dwindling quickly. Consequently, there is great motivation for unethical businesses to search for illegal suppliers. These are not hard to find: the Montreal Protocol allows for certain nonindustrialized countries to keep manufacturing Freon until 2010; a quick Internet search reveals several Chinese companies willing to ship Freon, complete with false "recycled" certificates; Mexico also produces Freon legally for roughly two dollars per pound; in the United States a pound can fetch 10 times that amount, making smuggling from Mexico very lucrative. It is not surprising that Freon ranks second only to cocaine as an illegal import.

Most Freon smuggling is the work of the Russian Mafia. Although, as an industrialized country, Russia was supposed to have phased out the manufacture of CFCs by 1996, it still has at least seven factories producing the chemical in defiance of international law. Russian spray cans still have Freon as a propellant, and there has been no significant move to introduce replacement refrigerants. These days, the Russians have bigger problems to deal with than the deteriorating ozone layer.

The Russian Mafia has recognized the economic potential of diverting Freon to the West, and it has been smuggling some thirty thousand tons every year into Western Europe and North America. Sometimes the containers they use are mislabeled as legal cooling agents; sometimes the CFCs are hidden in cylinders inside larger cylinders of a legal gas. Detecting the contraband is not an easy task. Usually, inspectors just check container pressure — a cylinder of Freon would have a different pressure from that of one of the legal refrigerant gases. But many of the smugglers are chemically astute, and they have figured out that they can add inert nitrogen gas to Freon, thereby raising the pressure to match that of a legal substance. In an effort to keep one step ahead of the Mafia, American inspectors have been armed with devices that attach to a cylinder's vent and can identify the cylinder's contents by measuring the extent to which the gas it contains absorbs specific wavelengths of infrared light.

The World Bank has also called on Western countries to reduce smuggling and help heal the ozone hole by donating forty to fifty million dollars to help Russian Freon factories switch to alternate products. So far, only about thirteen million has been pledged, probably to the relief of the Russian Mafia. Maybe one of the reasons the funds are not forthcoming is that some politicians realize ozone holes may be easier to deal with than bullet holes.

MENTIONING THE UNMENTIONABLES

As instructed, I lit a candle, blew through the panties I had stretched on an embroidery ring, and attempted to extinguish the flame. Nothing kinky about this procedure — I was just following the manufacturer's directions to prove that its line of polypropylene underwear allows the body to "breathe" better than underwear made from cotton or nylon.

Indeed, the candle was more easily extinguished by blowing through the polypropylene fabric than through the cotton or nylon. Then, as per the instructions, I soaked the garments in water and measured the time it took for each one to dry. Once again, the polypropylene fabric was the clear winner. What was all this supposed to prove? That underwear made from polypropylene, a fiber that had been used in making carpets, is actually "healthwear." The demos were designed to show that the polypropylene panties allow air to circulate and carry moisture from the inside to the outside, where it can evaporate. This, supposedly, makes the wearer less susceptible to feminine yeast infections, which, if we are to believe recent television advertisements, are the modern equivalent of the plague.

The health benefits of the "moisture-wicking" effect remain to be demonstrated, but it is undoubtedly an important property when it comes to keeping warm. Keeping dry is really the key to keeping warm, and it is therefore not surprising that polypropylene has met with success as a fiber for making long johns. Just one warning, though: polypropylene has a very low melting point; this is not a product for anyone who feels compelled to iron their underwear.

There are several varieties of thermal underwear on the market these days, and the manufacturers of most of them hype the fact that these garments allow moisture to pass through. One ad that caught my eye recently was for "Thermaskin" — "H_2O is attracted to Thermaskin like ants to a picnic. Our Constant Comfort process separates the H_2 from the O, making evaporation take place much faster." Does Thermaskin generate flammable hydrogen gas? Will this underwear blow up like the Hindenburg? Surely, the copywriter responsible for this needs a refresher course in chemistry: evaporation has nothing to do with water breaking down into its components;

it is merely a process of liquid water being converted into water vapor. There is really no need to worry consumers with the prospect of exploding undies.

Ascribing health benefits to undergarments is nothing new. Way back in the 1800s, Dr. Gustav Jaeger, a professor of physiology, initiated the Wool Movement. He advocated underwear made of coarse, porous wool, which allows the skin to breathe while keeping the body warm (supposedly, Pasteur's newly discovered microbes would lead to illness more easily if the body was chilled), but no medical consequences to people running around in wool bloomers — other than chafed skin — were ever recorded.

These days, the trend seems to be to employ underwear as a weapon in a direct attack on troublesome microorganisms. The British textile company Courtaulds has found a way to impregnate acrylic fibers with an antimicrobial compound known as irgasan. This disinfectant slowly rubs off onto the body and can prevent any disturbing odors that may be produced by the action of fungi and bacteria on compounds found in sweat and urine. There is even hope that this technology will be effective against candida albicans, the microbe responsible for feminine yeast infections.

Irgasan, so far, has checked out as a safe substance and is already extensively used in the production of mouthwashes, creams, and toothpastes. Courtaulds envisions employing it in towels and bedding, as well as in socks to combat athlete's foot. If these ventures are successful, underwear is sure to follow; a prototype of antimicrobial boxer shorts already exists.

Ah yes, boxer shorts. Who would have thought that they could involve chemistry? And biology? And controversy? In recent years, many men have switched to the more confining Jockey shorts, eliciting the concern of some scientists who have suggested that the shorts increase the temperature of men's private parts and may be associated with declining sperm counts and fertility rates.

Although there is no evidence for the "Jockey thing," it is well known that sperm production is temperature sensitive. A single 20-minute session in a sauna, for example, can reduce a man's sperm count for days. A sperm bank in Los Angeles has actually found that contributions made during the winter months have a higher sperm concentration and show increased sperm motility, so we should not be surprised that some scientists have even claimed that dinosaurs became extinct because they carried their testicles internally — as the earth's temperature increased, the creatures became sterile.

We don't exactly know what this dinosaur speculation means in human terms, but we do know that specially designed underwear equipped with "testicle coolers" (the designers' term, not mine) has shown some success as a fertility aid. We may also note that kilt-wearing Scots, who, according to tradition, let it all hang out, tend to have large families. British researchers, perhaps looking at the Scottish model, are presently conducting experiments with "scrotal slit" underwear, which provide some of the comfort of Jockeys along with appropriate cooling. In the meantime, Japanese scientists have found that men who watch X-rated videos more than double their sperm counts; furthermore, their sperm are more active, swimming faster towards their targets. Maybe video stores should have specials for Jockey-short wearers.

And just when you think you've explored the most zany aspects of underwear chemistry, along come underpants that release a compound that supposedly stimulates sexual interest. A company in Kanebo, Japan, has found a way to impregnate fabric with tiny capsules containing a substance found in male underarm secretions. This alleged aphrodisiac is released when friction breaks the capsules (the company makes no comment on how that friction is to be generated). When this product finally comes on the market we'll find out if it really does have the right chemistry.

SOME MAGICAL CHEMISTRY

Halloween is a time when witches fly, black cats prowl, and vampires rise from their graves. But it is also a time for magic. No, not supernatural charms, spells, or incantations, but the time-honored practice of prestidigitation. The last day of October is not only Halloween, but it is also International Magic Day, chosen because on this date in 1926 the most famous magician who ever lived, Harry Houdini, the man no chains or cuffs could hold, quietly slipped out of this world while a patient in a Detroit hospital.

Magicians are really scientists. Of course, they pretend not to be: they amaze their audiences with effects that appear to defy the laws of nature — but "appear" is the key word. What is really happening is quite different from what the observer thinks is happening. Magicians actually make use of very down-to-earth scientific principles to render the "impossible" possible. One of Houdini's favorite tricks was to produce lit candles from his various pockets. He performed this "miracle" by imbedding a match in the wax beside a candle's wick and wrapping a piece of sandpaper, clipped to the inside of his pocket, around the taper. When he pulled the candle rapidly, the match and the wick were ignited.

Ehrich Weiss, later to become Harry Houdini, was born in Budapest, Hungary, in 1874, but he emigrated with his family to Appleton, Wisconsin, while he was still very young. (Harry himself always claimed to have been born in America.) Legend has it that young Ehrich never cried — he lay silently in his crib, his eyes darting back and forth with curiosity. When, as a youngster, he went to a circus and saw a magician pull rabbits out of a hat as well as cut off and restore a man's arm, his destiny was sealed. He was going to be a magician.

He borrowed the name of the celebrated French magician Jean-Eugene Robert-Houdin, transforming himself into Harry Houdini, but he had no money to purchase professional supplies, so he had to rely on effects that could be carried out by sleight of hand or by using readily available materials, such as the "tincture of iron," then used to make inks. Houdini was developing an interest in chemistry. He learned that iron chloride reacted with a compound extracted from oak trees, known as tannic acid. The result was a dark, inky solution. He also discovered that the iron–tannic acid complex responsible for the color could be decomposed with the aid of oxalic acid, so he devised a clever trick and wove an ingenious story around it.

An elegantly attired Houdini strode out onto the stage, spoke of certain towns in America where the consumption of alcoholic beverages was prohibited, and went on to tell the tale of a clever chap who was able to outwit the inspectors who prowled the countryside by producing wine in a magical fashion. As he recounted the story, Houdini reenacted the miracle. He poured water from a pitcher into a glass, turning it into "wine." Pouring the contents of the glass back into the pitcher, he created a pitcherful of wine, but then he heard the inspectors knocking at the door. Quickly, he poured the wine back into a glass, whereupon it changed into water again. By adding this water to the pitcher, he then turned all of the wine back into water. The audience cheered as Houdini triumphantly held the pitcher of water aloft. The iron chloride, tannic acid, and oxalic acid had done their magic.

Houdini had diverse interests. He was as famous for his escape tricks as he was for his magic. As a child, he had loved his mother's apple pies so much that he would gobble them up as soon as they were cool enough to eat. Finally, Mrs. Weiss took to locking her pies up in a cupboard, but this did not deter young Ehrich. He quickly learned how to spring the lock, and this triggered a lifetime fascination with locks, handcuffs, and straitjackets. He became the man no lock could restrain.

He was also a superb physical specimen — he had to be in excellent condition to perform feats such as the famed Chinese-water-torture cell escape. He prided himself on having stomach muscles so strong that he could withstand any punch, and this proved to be his undoing. On October 19, 1926, Houdini gave a lecture at McGill University in Montreal on the techniques that fake mediums employ to convince their clients that they can communicate with the spirit world. Familiar with all the charlatans' tricks, taking delight in exposing them, Houdini often bragged that spiritualists would declare a national holiday when he left this world. Cleverly, he added that they could do this in safety because he couldn't get back at them from "the other side," but he also made it clear that if anyone could communicate from beyond the grave, it was he. The magician had a penchant for accomplishing "firsts."

At the McGill lecture, Houdini met young Sam Smiley, an arts student who had sketched a portrait of the magician during the presentation. Houdini invited Smiley to his dressing room at the Princess Theater, where the young artist could produce a more formal drawing. While Smiley was thus engaged, a divinity student called Gordon Whitehead came by with a book for Houdini. Whitehead asked Houdini if it was true that he could withstand any blow to the stomach. The magician affirmed that he could, but the blows came before he'd a chance to prepare himself. Less than two weeks later, the greatest magician of them all was dead from a ruptured appendix.

Numerous seances have been held on the anniversary of Houdini's death, but no messages have ever been received. The man who had cheated chains, prison cells, and handcuffs never managed to perform the ultimate escape.

DOWN THE HATCH

But It's Natural!

The word "natural" sells. Just plaster phrases like "natural goodness," "natural flavor," or "natural vitamins" on a label and sales soar, because many people assume that natural substances are somehow superior to synthetic ones, and they're willing to pay more for the perceived benefits. Consumers also tend to think that the creation of natural substances does not involve any kind of a manufacturing process. They are wrong on both counts.

Equating "natural" with "safe" and "synthetic" with "dangerous" is one of the greatest of all scientific fallacies. A little reflection quickly reveals that nature is not benign. Toxins produced by bacteria in food are perfectly natural, but they can be deadly. One of the most potent cancer-causing substances that exists is aflatoxin, the product of mold. Ricin, a protein found in castor beans, is probably the most poisonous chemical that has ever been isolated. One bite of the *amanita phalloides* mushroom can be lethal. Naturally occurring cyanide in cassava can kill. The HIV virus was not created by man. Natural sunlight can cause skin cancer and toxic algae can poison fish and the people who eat them. Poison ivy and stinging nettle can provide extremely unpleasant natural experiences. And we won't even speak of bee stings, scorpion bites, or snake bites.

Nevertheless, the myth persists that somehow substances produced by nature are superior to those fabricated in the laboratory. Natural vitamin C, extracted from rose hips, commands a far higher price than vitamin C made in the laboratory from glucose, even though the two are identical; their molecular structures are the same, and there is no way to distinguish the two substances. Natural vanilla flavor, extracted from the vanilla bean, is far more expensive than its synthetic counterpart, which can be made from pulp-and-paper waste. It may not sound very appetizing, but the synthetic compound, called vanillin, is identical to the vanillin found in the vanilla bean. Admittedly, the artificial version doesn't taste exactly the same as the natural, but that is because the natural flavoring is less pure — it contains a number of other compounds that occur in the vanilla bean in addition to vanillin.

To suppliers, the appeal of natural vanillin is that people are willing to pay a lot more for it. To many consumers, the appeal is the supposed health benefits of a natural product, and so the world supply of natural vanillin is not enough to meet the world demand. Madagascar, Tahiti, and Indonesia are the globe's leading growers of vanilla, but they cannot produce enough beans. The flavor industry has responded to this situation by stepping in and manufacturing synthetic vanillin, but the profit margin is considerably less than it would be if manufacturers were able to come up with a flavor that could be labeled as "natural vanilla flavoring."

How, then, can we get natural vanilla without growing vanilla beans? It may be possible to do so through a process usually referred to as "biotransformation." Biotransformations are reactions that make use of natural catalysts called enzymes. In fact, one of the oldest chemical reactions we know of is a biotransformation: the conversion of sugar to alcohol. Enzymes present in yeast do this job very well. We've been making wine and beer by means of such biotransformation for thousands of years. Molds and fungi are also great sources of enzymes; the flavor of Brie cheese, for example, is the result of the various compounds produced when enzymes that are present in the mold applied to the surface of the cheese react with the milk fats and milk proteins.

The microbial world provides us with a potential army of enzymes capable of carrying out a large variety of biotransformations. And today there is tremendous interest in finding specific microbes to do an enzymatic job, because the product of the reaction can still be labeled "natural." Here is an interesting example. One of the prime flavors in apples is a chemical called malic acid. This chemical is also responsible for the tartness of apples. It could, in theory, be extracted from apples, concentrated, and labeled "natural apple flavor." It could then be used in the making of apple-flavored foods, which could be described as containing natural apple flavor. It could also be used as an additive to adjust the acidity of processed foods, which could still be described as natural. But the extraction of malic acid from apples would be impractical and very expensive.

On the other hand, a specific microbe has been found that can create malic acid from fumaric acid, which is found in many plants. But we don't need fumaric acid from plants because a fungus called *rhizopus nigricans* can make it from glucose, and glucose can be made from starch by the mold *aspergillus niger*. Starch is plentiful and cheap. So the point is that we can make malic acid through a series of biotransformations that involve a great deal of technology and complex equipment, but the product can still be labeled as natural because the transformations are carried out by naturally occurring microorganisms.

Malic acid can also be made very cheaply by utilizing standard chemical techniques, but even though the malic acid created this way is exactly the same as the natural substance, by law it could not be labeled as natural, and so it would fetch only a fraction of the price. Now you can appreciate why flavor manufacturers are searching for microbes capable of producing vanillin from some common raw material such as starch: if they find one, they will be able to sell their product as natural, even though it will not taste like natural vanilla (because, as I mentioned earlier, natural vanilla contains several other compounds that contribute to its overall flavor). We are therefore left with this potential situation: a substance, say vanillin, will be available labeled either as natural or artificial depending on the process used to make it. The consumer will, of course, be charged far more for the natural version, and its suppliers' profits will be far greater. It clearly pays to know some chemistry, whether you are a producer or a consumer.

While biotransformations have an obvious commercial value in making products more marketable, they also have great scientific potential in making compounds that would be difficult to synthesize by standard chemical methods, or in producing them more cheaply. Consider the following example. One of the main flavor compounds in grapefruit is nootkatone, which is found in the fruit in very small amounts and is hard to extract. That's why the natural compound sells for ten thousand dollars a kilogram; a synthetic version made from valencene, a substance found in orange oil, is available for less but would be hard to produce on a large scale, and large-scale production of nootkatone has real commercial appeal. Recent studies have shown that grapefruit juice has the ability to increase the potency of several medications — such as cyclosporin, used to reduce the instance of post-transplant organ rejection; or lovastatin, used to lower blood cholesterol levels. Nootkatone is the substance that is most likely responsible for this effect. In the future, it may be possible to prescribe smaller doses of such medications if nootkatone is added to them. Side effects would also be reduced, and since cyclosporin is extracted from a fungus, and nootkatone is made by biotransformation, this high-tech pill could legally be labeled as "all natural."

While the term "natural" is often used in a meaningless or misleading fashion, one does have to admire the ingenuity of some of the marketers who employ it to describe their products. A hair-conditioner label brandishing the phrase "all natural ingredients" lists dimethicone as one of those ingredients. This compound is familiar to many as a "silicone." It's an excellent conditioner, but hardly natural. The manufacturer, when pressed, explains that silicones can be considered natural because they are made from sand, a decidedly natural substance. So what if a good dose of high tech and a few chemicals are needed to carry out the transformation? Who ever let a few little details like that get in the way of a clever marketing strategy?

And what about the case of "Natural Lemon Crème Pie," which contains sodium propionate, sodium benzoate, and artificial color? The presence of these preservatives and the color is not a cause for concern, but surely it should preclude the "natural" designation. "No," says a spokesman for the manufacturer: the label is justified. "Natural," he says, describes not the pie, but the lemon flavor. I suppose that they could have made the pie with artificial lemon flavor, but out of an admirable concern for the public welfare the company chose instead to use natural lemon flavoring. Then there is the television ad for a laxative that "works naturally, not chemically." And the carbonated bottled water that contains "natural carbon dioxide." We certainly wouldn't want any of that nasty synthetic CO_2 polluting our bodies. Such nonsense is enough to drive a man to drink — but what should he imbibe? Natural spring water, of course. Never mind that it may contain hydrogen sulfide or arsenic. They may be toxins, but so what? They're natural.

LIVING CAN BE FATAL

Chemistry was Denham Harman's favorite subject. He was absolutely captivated by chemical reactions, especially those that first produced extremely reactive intermediates and then went on to form a large number of products. "Could this also happen in living systems?" he wondered. "Might there be an explanation here for disease processes?" Harman decided that the way to find out was to mesh his love of chemistry with an understanding of biology, so he went to medical school. He now had all the tools for sound research.

In the 1950s, Dr. Harman began to study the effects of radiation on mice, and he noted that the animals exposed to it aged more quickly and died prematurely. This could be expected, he theorized, if the radiation sparked the production of some highly reactive and destructive chemical species capable of wreaking havoc in tissues. At the time, this idea did not generate much interest, but soon the scientific world was vigorously debating the existence of Harman's rogue "free radicals" — and they weren't talking about liberated political dissidents.

These capricious molecular free radicals, Harman suggested, could be tamed. In fact, the food-processing industry was already doing this by adding certain preservatives to its products. Butylated hydroxy toluene, or BHT, was being widely used to prevent fat from going rancid. It worked, Harman said, because the rancidification of fat was a reaction mediated by free radicals, which were neutralized by BHT. To make his point, he added BHT to the diet of rats and saw that their life expectancy increased. Still, there being no practical evidence of free-radical mischief, the scientific community remained skeptical — until 1969. That year, an enzyme called superoxide dismutase (SOD) was discovered in cells. Apparently, its sole function was the destruction of a type of free radical known as superoxide. Why would this enzyme be present in all cells if free radicals did not represent a danger?

Research now accelerated. A newly developed instrumental technique known as electron spin resonance (ESR) demonstrated that free radicals did indeed exist in biological systems and soon they were being linked with heart disease and cancer, the major causes of early death. But what was the connection?

Scientifically speaking, "life" is just the result of all the simultaneously occurring chemical reactions in our bodies. Some of these reactions are involved in building muscle, some in destroying invading bacteria, some in growing hair, some in synthesizing sex hormones, and some in fostering our thought patterns. Then there are the reactions that generate the energy required to fuel all of the other processes — and therein lies the problem. Life is actually deadly. The reactions that produce the energy allowing us to carry on with our lives also create the insidious free radicals that may hasten our demise.

Just as a furnace burns oil for energy, our cells burn glucose; and for this combustion process our cells, like the furnace, need a constant supply of oxygen. The oxygen is absorbed through the lungs, incorporated into hemoglobin molecules in red blood cells, and delivered via the bloodstream to the trillions of microscopic furnaces we call cells. Here, the oxygen begins the process of deriving energy from glucose by breaking the chemical bonds that hold the molecule together. Chemical bonds are nothing more than a pair of electrons between the atoms that are joined together. Oxygen strips an electron from glucose and thereby disrupts the "glue" holding the molecule together. This, in turn, begins the cascade of events that ultimately converts glucose into carbon dioxide and water, releasing energy in the process.

Now the oxygen is stuck with an extra electron — an unfulfilled chemical bond. It has become a free radical, desperately searching for some molecule with which it can react to satisfy its craving for electrons, and there are plenty of candidates in that cellular chemical stew. Fats and proteins, for example, are readily oxidized; so is deoxyribonucleic acid, or DNA, the master control molecule of life. Unfortunately, when these compounds sacrifice themselves to appease the reactive oxygen, they are altered in such a way that they can no longer fulfill their function. Certain proteins lose their elasticity and our skin wrinkles. Other proteins, responsible for carrying cholesterol around the bloodstream, are oxidized to a form that damages arteries, perhaps causing coronary disease. The fats in cell membranes may become rancid, shortening the cells' lives. Worst of all, damage to the DNA molecules that control the multiplication of cells can cause cancer. It is not a pretty picture.

We're still not finished. Depressingly, there are yet other ways in which free radicals can form. Certain white blood cells produce them as weapons against invading organisms, but healthy tissues may suffer from friendly fire and inflammation results. Sunlight, X-rays, and environmental pollutants such as ozone can also trigger free-radical formation. Given this dark scenario, how is it that we actually survive as long as we do? Because our bodies have some remarkable defense weapons: the antioxidants. We do not take the free-radical onslaught lying down. Cells respond by synthesizing enzymes like superoxide dismutase and glutathione peroxidase, which destroy free radicals. And then there are the all important, highly publicized, dietary antioxidants. Vitamin C, vitamin E, and beta-carotene all react with free radicals, thereby sparing other molecules from attack, as elegantly shown by Professor Keith Ingold at the University of Ottawa. We are not only talking esoteric theory: study upon study now suggests that we can improve our prospects for health by increasing our intake of foods rich in these antioxidants.

The lingering question in the chemist's mind, however, is whether we have oversimplified the whole antioxidation scenario with our undue emphasis on vitamin E, vitamin C, and beta-carotene. After all, food is an incredible collage of chemicals. Isn't it possible, or indeed even likely, that the beneficial effects of fruit and vegetable consumption are the result of cohesive teamwork among many antioxidants? Most of the over two hundred studies that have demonstrated the benefits of antioxidant consumption have been based on fruit and vegetable, rather than supplement, intake. These foods have numerous components — other than the superstar vitamins — that can contribute to antioxidant activity. A Cornell University study, for example, clearly showed that vitamin C as an integral part of fruit juice is more effective at reducing the formation of carcinogenic nitrosamines in the body than vitamin C as a supplement: it seems that chlorogenic acid in the juice boosts vitamin C activity. So what are we to do?

When it comes to one particular point there is no argument. We should be eating a variety of fruits and vegetables every day — at least five servings. As the ancient saying goes, the more color on the plate, the better. But there is a second point of fact: only a minority of North Americans consume the suggested five servings a day, and, for various social and economic reasons, no amount of cajoling is going to change this dramatically. Furthermore, even diets rich in fruits and vegetables may not include the optimal amount of vitamin E. So, although many claims about antioxidants may be exaggerated, current research supports daily supplementation with two hundred milligrams of vitamin C and two hundred to four hundred international units of vitamin E. While it is unrealistic to expect these nutrients to compensate for the effects of an unhealthy lifestyle, their benefits outweigh any risk.

Researchers may eventually show that other antioxidants may also be appropriate as supplements — perhaps even BHT to slow down aging, which, after all, is just a slow rotting process. Those wary of the prospect of BHT pills can just gulp some more cereal, but check the label: make sure the product is preserved with BHT.

An Eye-Catching Story

The exploits of the Royal Air Force during the Battle of Britain have become legendary. Why were British pilots so successful in downing the Luftwaffe bombers? According to the Air Ministry, they gained their advantage by dining on carrots. This explanation sounded reasonable, even to German military intelligence, as scientists had already established that vitamin A deficiency could cause night blindness. Furthermore, it was known that beta-carotene, one of the orange-colored carotenoids found in carrots, could be converted by the body into vitamin A. If carrots made the British see better in the dark, surely they would do the same for the Germans. So the Luftwaffe ordered its pilots to eat carrots before their missions, but no matter how many carrots they devoured, British air superiority was maintained. This, as it turns out, was because the Royal Air Force's success actually had nothing to do with carrots.

The pilots' uncanny night vision was not attributable to vitamin A, but rather to a new invention called radar. The southern and eastern coasts of England had been lined with radar installations that could pinpoint the approaching German bombers for the RAF. The Air Ministry had, in fact, cooked up the carrot story and fed it to German intelligence in order to send them searching for carrots instead of radar antennae. Carrots may not have improved the pilots' eyesight, but recent research has revealed that beta-carotene does play a significant role in health maintenance. This is probably due to its ability to act as an antioxidant and neutralize the troublesome free radicals that have been linked with cancer and heart disease. A Johns Hopkins University study of more than 25,000 people who had their blood sampled over a 10-year period supports this notion. Subjects with low beta-carotene levels had four times the rate of a certain form of lung cancer. The Western Electric study in Chicago, which monitored the health status of 2,107 workers for 19 years, also found that the incidence of lung cancer in smokers who had low carotene intakes was seven times greater than in smokers who ate a lot of carotene-rich foods. At the Albert Einstein College of Medicine in New York, researchers found a threefold greater risk of cervical cancer in women with low carotene intake.

There are also interesting links between beta-carotene intake and heart disease. The 22,000 doctors enrolled in the Physicians' Health Study were asked to take either a 50-milligram beta-carotene tablet or a placebo every second day. While no significant differences were noted in cancer rates, supplements cut the heart-attack risk in half among the subjects who had signs of heart disease when they entered the study.

The largest longterm study of women in the world is the Nurses' Health Study, coordinated by Harvard Medical School. It involves over eighty thousand nurses who, since 1980, have periodically answered detailed dietary questionnaires and have had their blood sampled randomly for vitamins and beta-carotene. Over the period of the study, women consuming a diet containing more than 15 to 20 milligrams of beta-carotene daily had a 40 percent reduced risk of stroke and a 22 percent reduced risk of heart attack compared with women taking less than six milligrams. Of one thousand women who had angina, the highest carotene consumers had an 80 percent reduced risk of heart attack.

These studies were highly publicized in the lay press, and many people started popping beta-carotene supplements, but the bandwagon screeched to a halt in 1994 with the revelation that in a Finnish study smokers who took beta-carotene supplements were actually more likely to develop lung cancer. This was a shocker, since researchers had expected that smokers — who, of course, are at greater risk for lung cancer — would derive the greatest benefit from beta-carotene. Critics tried to pass off these findings as anomalous, but they were quieted when an American study of smokers also showed an almost 30 percent increase in lung cancer among subjects taking 30 milligrams of beta-carotene daily in supplement form. What was going on here? Researchers at Tufts University tried, literally, to ferret out the answer. They fed high doses of beta-carotene to the weasel-like animals, which metabolize the compound the same way humans do. Some of the ferrets also inhaled the equivalent of 30 cigarettes' worth of smoke a day for six months. The incidence of lung tumors increased, especially among the smoking ferrets. But this time analysis of the animals' blood suggested a solution to the paradox: at high levels, beta-carotene actually acts as an oxidant instead of an antioxidant.

Beta-carotene's antioxidant effect can be ascribed to the fact that it can neutralize free radicals by donating an electron. In the process, however, it itself becomes a free radical capable of tissue damage unless it is appeased by some other molecule from which it can snatch an electron. This is where vitamins E and C enter the picture. These compounds are adept at scavenging the carotene radical without generating dangerous species. Since smokers are known to have low blood levels of vitamin C, they can be expected to be at greater risk from beta-carotene supplements.

More evidence for the unusual behavior of beta-carotene comes from — of all things — chicken feed. Fat is commonly added to improve the efficiency of feed utilization. Unsaturated fat is better because it improves the nutritional profile of the final product, but unsaturated fats in the meat oxidize more easily than saturated fats, degrading taste and texture. Producers have experimented with fortifying the feed with vitamin E and beta-carotene to reduce oxidation. They have discovered that when beta-carotene is added, it behaves as an oxidant unless vitamin E is added as well. With sufficient vitamin E, however, beta-carotene exerts its expected antioxidant effect.

So what are we to do with this information? For the moment, we should probably lay off beta-carotene supplements, but we should certainly continue to eat foods that are rich in beta-carotene, because beta-carotene may require other food components in order to exercise its benefits. There is no recommended daily intake of beta-carotene, but a review of the literature reveals that we should strive for about 20 to 25 milligrams daily. To put this amount into perspective, a sweet potato has about fifteen milligrams; a carrot, twelve; half a cantaloupe, five; half a cup of spinach, four; and a spear of broccoli, two.

We've seen the benefits of beta-carotene — and talking about seeing, there is one final story. It has nothing to do with night blindness; it has to do with cataracts, the leading cause of blindness in the world. As we age, free-radical reactions cause the protein in the lens of the eye to clump and form opaque deposits that scatter light before it can get through to the retina. We call these deposits cataracts. Several recent studies have shown that a high intake of antioxidant nutrients, particularly carotenoids, is associated with a decreased risk of cataract formation.

Carrots may not have defeated the Luftwaffe, but they may help us win the war against cancer and heart disease. By reducing the risk of cataracts, they may even help us to see the future more clearly. There's surely a wealth of health in yellow and orange fruits and vegetables — and that's not chicken feed.

LET'S HAVE AN APPLE DAY

As an educator, I have a special fondness for apples, the traditional gift of student to teacher. This gift-giving practice may have something to do with the apple's biblical reputation as "the fruit of the tree of knowledge of good and evil," although the Bible never actually states that the fruit in question was an apple. We probably owe this interpretation to medieval artists who were more familiar with apples than with pomegranates or apricots, more likely candidates due to their prevalence in the Middle East.

Nevertheless, the role of the apple in the story of Adam and Eve has become so widely accepted that even the botanical name for the fruit, *malus pumila*, derives from the Latin word for evil. A lover of apples can then be called a "malophile," and a big lover of apples is a "megamalophile." (Or would that be a lover of the Big Apple?) The unfortunate experience of Snow White aside, there is no such thing as an "evil apple" — not even the ones that are grown with the aid of those dreaded "chemicals." Yes, fungicides, pesticides, and plant-growth regulators are sometimes used by apple growers, but a rational, scientific examination of the effects of these substances arouses very little concern.

Farmers would prefer not to use agricultural chemicals. First of all, they are expensive: a six-hundred-gram bag of pyridaben, a compound that controls mites in the orchard, costs hundreds of dollars. Second, the greatest risk posed by these substances is to the farmer, who has to prepare solutions for spraying (modern technology has, however, reduced this hazard; many agricultural chemicals are now packaged in water-soluble bags made of polyvinyl alcohol). Third, spraying is most effective when done after dark — usually around 4 A.M. — and a farmer's day is already long enough.

Clearly, apple growers spray out of necessity, to produce a financially viable crop. Apple scab, the European sawfly, caterpillars, fruit flies, maggots, mites, and other assorted fungi and insects are just too potent a threat. Captan, a fungicide, can halt apple scab, which causes fruit to discolor and fall prematurely. Insecticides like phosmet can ensure that we don't bite into an apple and find half a worm inside.

These chemicals are not applied in a haphazard manner. For example, fake apples covered with glue are hung in the orchard to collect insects, and only when it is apparent that there is an infestation of, say, apple maggot fly, is an insecticide applied. The need for fungicides is determined by careful monitoring of wetness and temperature. Plant-growth regulators, like naphthalene acetic acid, can strengthen the fruit's stem and allow proper ripening, but if too much is used, the apple stays on the tree too long and softens. Appropriate use of calcium chloride yields firm apples, zinc compounds increase yield, and various nitrates provide trees with essential nitrogen. Carbaryl thins out the fruit to ensure adequate size. This is clearly a complex business.

The uses and effects of all these agricultural chemicals have been studied extensively: the time it takes for the compounds to break down in the environment is well known, and permissible residue levels have been determined with significant built-in safety factors. One is far more likely to be struck down by lightning while picking apples than to be adversely affected by chemical residues.

Of course, apples can be grown without the support of agricultural chemicals, but crop yields are dramatically reduced. This makes the fruit more expensive, decreases consumption, and deprives people of important potential health benefits. These benefits, which have been recognized for centuries, are now being corroborated by modern science. Englishmen came to the New World armed with apple seeds because they knew that water carries disease whereas apple cider does not, and so cider became one of the most popular beverages in the new colonies. Soon, apples, partially through the romanticized work of John Chapman, better known as Johnny Appleseed, took their place beside motherhood in American lore. Apple trees were to be found everywhere, and the land reverberated with the dictum "An apple a day keeps the doctor away" (though it may actually take three). This dictum is supported by our current extensive knowledge of the chemical composition of apples and by the properties of their specific constituents as elucidated by researchers.

Pectin, for example, lowers cholesterol. Adding two to three apples a day to your diet may lower your blood cholesterol level by 11 percent. Pectin also controls blood sugar, and in test animals it has lowered the rate of colon cancer by 50 percent. Chlorogenic acid and other phenols can neutralize carcinogens, at least in test-tube studies. Apples are even rich in boron, an element that may reduce the symptoms of osteoarthritis. A study conducted at Michigan State University has shown that students who eat the most apples have the fewest upper-respiratory-tract infections, while Dutch researchers have demonstrated that flavonoids in apples can significantly reduce heart-attack risk. And which apple has the highest concentration of flavonoids? Jonagold, a cross between Jonathan and Golden Delicious.

The wonders of apples just keep accumulating: munching on apples reduces the risk of cavities; scientists at the Taste Research Foundation of Chicago have even found that apple smell can reduce anxiety. So anyone still worried about chemical residues on apples should just take a good whiff of an apple and then eat one. Or, better yet, eat three. With all this wonderful news about apples, it is no wonder that in England they actually celebrate Apple Day every October 21st. I think this is a great idea, and I propose that we all adopt the tradition. So, next October, get out there and wish everyone a happy Apple Day!

EGGSPERTISE

"A box without hinges, key, or lid / Yet golden treasure inside is hid." That's how J.R.R. Tolkien described the egg. It is, indeed, a fascinating little box, although these days not everyone agrees that its golden contents amount to a treasure. They're worried about cholesterol, and there really is a good dose of it in an egg yolk — two to three hundred milligrams — but there is also a good dose of misunderstanding about how this affects our blood cholesterol levels.

The amount of cholesterol that circulates in our blood is much more a function of the saturated-fat content of our diet than of its cholesterol content. Studies show that most people's blood cholesterol is increased, at most, by 2.5 percent for each egg eaten, a little more for those who also have elevated triglycerides. This could be significant for someone eating two eggs every day, but it doesn't mean much for the three-to-five-egg-a-week consumer. There are even some lucky people whose blood cholesterol seems totally unaffected by diet: the *New England Journal of Medicine* has reported on the case of an 88-year-old gentleman who for 15 years ate up to 25 soft-boiled eggs every day while maintaining a normal cholesterol level. But before anyone uses this example to try to justify their own egg gluttony, I should point out that studies have indicated that people who consume relatively large quantities of cholesterol are more likely to die of heart attacks, regardless of blood cholesterol levels. Some researchers, therefore, argue that dietary cholesterol can somehow do damage even without increasing the amount of cholesterol in the blood.

Obviously, the effect of eggs on the heart is debatable. Egg-o-philes and egg-o-phobes can both marshal studies to advance their views, and that's because it's difficult to get hard data on the effects of egg consumption — just about as difficult as making a perfect hard-boiled egg. But at least in this case the science isn't soft; with some chemical knowledge we can create an egg that doesn't crack during cooking, that peels easily, that doesn't have a flat bottom, and that isn't tainted with that revolting greenish color around the yolk.

First of all, we need to understand why an egg hardens when cooked. This is simply a matter of protein chemistry. A raw egg is mostly water in which protein molecules, along with some fat and cholesterol, are suspended. The proteins, long chains of amino acids, are coiled up like little balls of string, and they interact only minimally. Heat causes these molecules to uncoil, exposing sites on their surfaces where links to other protein molecules can be forged — it is as if the strings straighten out and then intertwine. This microscopic clustering is manifested as macroscopic hardness. The molecular clusters also reflect more light, so the cooked egg loses its transparency. If heating goes on too long, water molecules that have become trapped in the protein lattice are squeezed out, leading to an even tighter protein structure and a rubbery texture. No one likes an overcooked egg.

What about the dreaded flat bottom? This can occur because a raw egg does not completely fill its shell, thereby allowing for a little air pocket that actually provides a chick with its first breath. During the cooking process, if this air is not permitted to escape before the white hardens, the egg will develop a flat bottom. Older eggs are more prone to this because they have lost some moisture and therefore have a bigger airspace. As the air is heated, it expands and begins to escape through the porous shell — this is often evidenced by a telltale column of bubbles rising from an egg immersed in hot water.

If the air expands too fast, however, it can crack the shell and torment the cook with white streamers. These form as the liquidy egg white, the albumen, spills out and coagulates in the hot water. Adding a little salt or lemon juice to the water can circumvent this problem, because, like heat, these reagents immediately cause the proteins to unfold, join together, and harden as the white oozes out; as a result, the crack is sealed. But it is better to prevent the calamity rather than try to fix the problem after it happens: just take a nail or a thumbtack and make a little hole in your egg prior to cooking; the escaping air will prevent pressure build-up and allow the egg white to flow into the space previously occupied by the air. No flat bottom now.

Still, there's more than one way to crack an egg. If a cold egg is placed in hot water, its shell begins to expand, and since the shell is not of a consistent thickness, some areas expand more than others — the resulting stress leads to a fracture. The solution is to place the egg in cold water, bring to a boil, reduce to a slow simmer for 10 minutes, then immerse in cold water and peel. If this sounds like too much trouble, you can seek out a farmer who treats his hens to carbonated water; this increases the carbonate concentration in their blood, resulting in stronger eggs. I'm told this is done in hot climates, where chickens pant a lot and exhale a great deal of carbon dioxide. Apparently, they prefer Perrier.

An eggshell can sustain a stress fracture while cooling, as well; some parts of the shell will contract faster than others. The young man who heated seven eggs in their shells for five minutes in the microwave found out about this the hard way. Somehow, the eggs survived the pressure build-up in the oven, but when the hungry man sat down to enjoy the fruits of his labor, six of the eggs exploded. He was severely burned about the face.

How easily an egg peels depends on its age. Fresh eggs tend to have a higher carbon-dioxide content, and some of this gas dissolves in the egg's moisture to form carbonic acid. As an egg ages, the carbon dioxide diffuses out and the contents become less acidic. This weakens the inner membrane that surrounds the egg and prevents it from sticking to the hardened white. Researchers have proven this by showing that fresh eggs exposed to ammonia vapor, which neutralizes acids, can be readily peeled. And how do you know if there is going to be a peeling problem? Just place the egg in a pot of cold water: a fresh egg has a small air pocket and will sink; an older egg will float.

Even a perfectly shaped and peeled hard-boiled egg can have an inner secret — the greenish outer yolk. Not to worry, it's just a small amount of benign iron sulfide. Egg yolk contains iron, which reacts with the hydrogen sulfide gas that forms when some of the proteins in the white of the egg decompose during heating. The older the egg, the more likely the formation of hydrogen sulfide. As the gas forms, its pressure increases due to the high temperature, and it migrates towards the cooler regions — namely the yolk — so the remedy for yolk discoloration is to immerse the cooked egg quickly in cold water, thereby providing an alternate region of low pressure to attract the hydrogen sulfide.

Let's put it all together. Start with an egg that's been refrigerated at least a week, make a hole in the bigger end with a thumbtack, immerse in salted cold water, bring to a boil, reduce to a simmer for 10 minutes, plunge into cold water, and peel. Don't use a microwave. You may end up with egg on your face.

CHINESE-RESTAURANT SYNDROME

"No MSG" proclaims the sign in the window of the Chinese restaurant, referring to the popular flavor enhancer. Why eliminate it? Don't people go to restaurants to be served tasty food? Sure they do, but they also go with the hope that they won't have to sacrifice their good health for enhanced flavor, and there is a common perception that this is exactly what happens when monosodium glutamate is thrown into the mix. This perception is generally incorrect.

Monosodium glutamate (MSG) may be the most maligned

and misunderstood of all food additives. Most people assume that MSG is one of those questionable substances unleashed upon the public by modern chemical technology. In fact, glutamic acid and its derivatives are widespread in nature. Even the human body contains glutamate, which is a neurotransmitter, or chemical communicator, for sensory nerves entering the central nervous system.

Oriental cooks have long flavored foods with seaweed and soy sauce, but it was not until 1908 that a Japanese researcher discovered the secret substance behind their effect: monosodium glutamate, a salt of a commonly occurring amino acid. Today, MSG is produced on a massive scale by a fermentation process in which beet sugar or corn syrup is converted to glutamic acid.

Glutamic acid is also found in a wide variety of foods. The ability of mushrooms and tomatoes to intensify the flavor of certain dishes depends on their high free-glutamic-acid content. Parmesan and Camembert cheeses also owe their characteristic taste in part to glutamate. Obviously, then, glutamic acid is prevalent in the foods that we eat on a daily basis. Indeed, calculations show that our intake of naturally occurring glutamic acid from foods is far greater than it is from the additive MSG. So why all the fuss over monosodium glutamate?

In 1968, an American physician of Chinese extraction, Dr. Ho Man Kwok, noted a tightness in his chest and jaw, a headache, and a burning sensation in the back of his neck a short time after eating a meal in a Chinese restaurant. In a letter to the *New England Journal of Medicine*, he coined the term "Chinese-restaurant syndrome" (CRS) for this grouping of symptoms.

The occurrence rate of Chinese-restaurant syndrome is a controversial issue. There is no doubt that some people do suffer from the condition, and that in some cases it can be induced by the intravenous administration of MSG. The subjective symptoms, however, are not consistent, and objective symptoms such as heart rate, blood pressure, and skin temperature are unaltered during an "attack." A much more serious, but fortunately very rare, response to MSG is an asthmatic attack. There is at least one recorded case of respiratory arrest — it happened to a woman within an hour of consuming a meal of wonton soup, almond chicken, and Szechuan beef. She had not eaten for 11 hours before the meal and was also known to be sensitive to sulfites. In this particular case, a double-blind challenge did reveal that MSG was the likely cause.

There have been other concerns as well. A 1969 study by University of Washington researcher John Olney concluded that massive doses of MSG destroy brain cells in mice. This finding led to baby-food manufacturers voluntarily eliminating MSG from their products. Nevertheless, the relevance of this study to humans is questionable, since a number of subsequent studies involving primates showed no effect upon subjects when they were injected with, or force-fed, MSG.

Only in one rare case has MSG been the apparent cause of seizures — in a young boy — and, of course, the fact that the seizures in question resolved when foods that contained MSG were eliminated from the sufferer's diet does not amount to conclusive proof of MSG's culpability. Numerous other food components were also eliminated when processed foods were removed from the boy's diet.

Fuel was added to the MSG fire in 1992 when the influential CBS program *60 Minutes* aired a segment about a woman who alleged that the failure to recognize MSG as the cause of her stomachaches led to unnecessary surgery and a mother who claimed that her son's hyperactivity and low grades were due to MSG. Then there was Olney, wrapped in a white lab coat, suggesting, without any convincing evidence, that MSG can cause brain damage in some people. The program was irresponsible, merely paying lip service to the massive amount of research carried out since 1968, research showing that while there may be some isolated, idiosyncratic reactions, MSG is no nutritional villain.

Jaded reports about MSG are not all generated by television producers eager for sensational stories. The National Organization Mobilized to Stop Glutamate, which may have exhausted its mental acuity coming up with a name that would create the acronym NOMSG, has released reams of information about the evils of MSG. Its members should have directed their efforts towards reading the scientific literature.

In 1992, the US Food and Drug Administration (FDA), prompted by public concern about MSG, asked an independent panel of scientists, the Federation of American Societies for Applied Biology (FASAB), to study the issue. A comprehensive 1995 report based on well-controlled double-blind studies concluded that MSG presents no problems when consumed at normal levels but that a large dose could cause a burning sensation, facial pressure, headache, drowsiness, and weakness in a very small percentage of people.

Significantly, there was a threshold effect: the symptoms I just named were noted only in people who had ingested over 2.5 grams of MSG at one sitting. It is possible to consume this much with some Chinese meals. Still, the researchers felt that it was unfair to saddle Chinese food with a pejorative label since other meals, such as spaghetti with tomato sauce and Parmesan cheese, could have equally high levels of glutamate, so they opted to rename Chinese-restaurant syndrome, dubbing it "MSG symptom complex."

A recent Canadian study has further underlined the safety of MSG. An investigation of 61 subjects who claimed to be MSG sensitive clearly showed that at less than 2.5 grams there was no difference between MSG and a placebo. The average North American consumption of the additive is 0.55 grams per day.

Scientific evidence clearly shows that MSG is not a scourge on society. Often, it serves as a scapegoat for any ill effect a person may experience after a meal — in fact, about 40 percent of the population report unpleasant symptoms after being tested with any food. Some do, however, react to MSG, primarily when exposed to high doses on an empty stomach. Symptoms vary, and they are generally transient, benign, and not reflected by objective measurements or by blood levels of glutamate. In rare cases, large amounts of MSG may trigger asthmatic attacks.

All this talk about flavor has made me hungry. I could go for some pizza. I know how great that combo of mushrooms, cheese, and tomato sauce will taste — must be all that natural glutamate. I just love that Italian-restaurant syndrome!

HOT DIGGETY DOG

I have a confession to make: I like hot dogs. In these days of nutritional correctness, saying this makes me feel as though I'm admitting to some criminal tendency. At the risk of riling people devoted to subsisting on alfalfa sprouts, algae, tofu, and diverse supplements, let me assure you that it is possible to indulge in hot dogs occasionally and still have a healthy diet. It is also possible to shun wieners and still have a diet that is a nutritional nightmare. Individual foods should not be vilified or deified; it is the overall diet that determines whether we are eating in a healthy or an unhealthy fashion. In any case, like it or not, sausages in various forms have been with us a long time and are destined to remain part of our nutritional culture for the simple reason that they taste very good.

People have been stuffing ground meat along with various spices and other ingredients into casings for thousands of years. Homer sang of sausages in *The Odyssey*, circa 850 B.C. The Romans traditionally made sausages from ground pork and pine nuts for the celebration of Lupercalia, a festival of eating, drinking, and wenching, and these sausages became so intimately connected to debauchery that Constantine, the first Christian emperor, actually banned them. Sausage bootlegging then became a profitable enterprise.

By the Middle Ages, hundreds of varieties of sausage had been developed. Many of these, like bologna, were named after the city where they were first made, but the most enduringly popular variety originated in the German city of Frankfurt. The frankfurter was made of cured meat and cooked with smoke. Legend has it that the sausage was introduced into North America in 1904 by Anton Ludwig Feuchtwanger, a Bavarian peddler who set up a booth at the St. Louis World's Fair. Since the sausages he sold were greasy and hot, he loaned his customers white gloves to wear while holding them, but so many people absconded with the gloves that he had to find another solution. His brother-in-law, a baker, came up with one: why not put the frankfurter in a bun?

Everyone wanted to try the new "Dachshund sausages," as the franks were quickly dubbed, because of their resemblance to these elongated canines. Soon the name evolved into "hot dog," and a North American staple was born. Today, hundreds of manufacturers compete to satisfy the North American craving for some sixty million franks a day.

Many of us, however, eat those hot dogs with a certain trepidation, because we're not quite sure what they contain. Otto von Bismarck, the celebrated German statesman, once remarked that the two things you don't want to see made are sausages and the law. Judging by some of the parliamentary behavior I've witnessed, he was right about the law, but sausages aren't really so scary — we can actually learn a lot about science by investigating how they are made.

No matter what you may have heard, there are no ears, snouts, or genitals in your hot dogs. They can be made from the edible parts of beef, veal, lamb, pork, or poultry; this may include tongue, heart, esophagus, and blood. If you find that hard to stomach, I probably shouldn't tell you that manufacturers also sometimes use stomach. Kosher hot dogs do not contain any of these delicacies; they are made from good-quality lean meat mixed with "plate trimmings," which is essentially a pseudonym for fat.

Whatever kind of hot dog you buy, it will be a product of the same basic process. The ingredients are finely chopped and blended into a smooth paste, which is then stuffed into a casing and cooked. The special flavor is created from a mixture of spices that includes garlic, pepper, paprika, smoke flavoring, and MSG. Vitamin C or its chemical cousin, sodium erythorbate, is also mixed in. Why vitamin C? Because it mitigates the action of the curing salt, which is added next. The curing salt is a mixture of about 98 percent regular salt and 2 percent sodium nitrite.

Nitrites are perhaps the most controversial hot-dog components. They add flavor and color, and they prevent the growth of the deadly *clostridium botulinum* bacteria, but they can also react with other substances in meat, called amines, to form nitrosamines. While these substances are carcinogenic in test animals, and probably in humans, the risk they pose is very small. The odd study has linked hot-dog consumption to some rare childhood cancers, but critics have pointed out that if hot dogs are indeed the culprit, they are only so when it comes to vitamin-deficient children — another reason to make sure your kids are taking their multivitamins.

In any case, food processors have greatly reduced their use of nitrites since the discovery that vitamin C potentiates the action of these chemicals; this means that less nitrite can be used if vitamin C is added to the mix. Studies have shown that the added vitamin C also reduces the chance of nitrosamine formation in the body. It is possible to make nitrite-free hot dogs, but these must be kept frozen.

If the nitrite issue isn't that significant, why shouldn't we feast freely on hot dogs? It's because of the fat: by law, the protein content of hot dogs must be at least 11 percent, but the fat content is not regulated. The average hot dog is 23 percent fat by weight. That's a lot — a T-bone steak is 12 percent fat by weight. The average hot dog contains about 10 to 15 grams of fat, most of it saturated, although poultry and veal franks contain somewhat less. This is substantial, considering that our daily fat intake should not routinely exceed 60 to 70 grams, but it's the fat in the hot dog that makes it taste so darn good.

Is it possible to have a low-fat hot dog? Well, the American company Hormel has come up with "97% Fat Free Franks," which only have 1.3 grams of fat each. Hydrolyzed vegetable protein has been used to replace some of the fat, and this is certainly a giant step in the right direction, especially when one considers that a panel of tasters found the Hormel product as tasty as regular hot dogs. Incidentally, the package label "100% beef" is nutritionally meaningless. This merely indicates that all the product's components, including the fat, are derived from cows, steers, or bulls. Bull meat is actually very flavorful, but because it is so fibrous it tends to be tough. However, when macerated in a blender, it makes for an ideal hot dog. And that's no bull.

Then there are tofu hot dogs. These are getting better, but they still seem to develop those revolting "warts" when grilled. For now, I'll still choose the occasional regular hot dog, especially if I can have a good ball game with it. Just to be on the safe side, I'll order it with sauerkraut — lots of vitamin C in there to take care of any nitrite problems.

It's Always Tea Time

I think we should be consuming more epigallocatechin-3-gallate. The term may twist a few tongues, but the stuff is raising more than a few eyebrows. That's because EGCG, as it is mercifully abbreviated, may offer significant protection against the two greatest plagues of our time: cancer and heart disease. And the good news is that to reap its benefits we don't have to swallow pills, drink strange concoctions, or eat bean sprouts; we just have to get into the habit of drinking tea.

This is not a hard habit to get into. Billions of people around the world drink tea regularly — it is the most widely consumed beverage after water. Legend has it that the first cup of tea was brewed accidentally in the court of the Chinese emperor Shen Nung almost five thousand years ago. The emperor's drinking water was routinely boiled, and some leaves from the burning branches under the pot drifted into the water. Drinking the brew anyway, the emperor noted that it not only quenched his thirst but it also decreased his urge to sleep. What was good enough for the emperor was good enough for the populace, and tea became the most popular drink in China.

The beverage was introduced into Europe in the seventeenth century, but it was not an immediate success. Tea was expensive, and many members of the clergy branded it a dangerous intoxicant. Sometimes their denunciations took spectacular forms: in England, the Reverend Dr. Hales demonstrated that if the tail of a suckling pig were dunked in a cup of tea, it would emerge hairless. Most people recognized the absurdity of this experiment, and despite the reverend's efforts, tea sales did not tail off.

Today, more tea is consumed than ever before. All tea — except the herbal varieties — is made from the leaves of the *camellia sinensis* plant. Exactly which leaves are picked determines the flavor of the beverage: orange pekoe tea, for example, is brewed from the second leaf next to the bud, pekoe from the third. Evidently, picking tea is not an easy task; neither is unraveling its chemical mysteries.

Like many natural products, tea contains hundreds of compounds. The most characteristic of these are often referred to as flavonoids or polyphenols. A subclass of the flavonoids, the catechins, to which EGCG belongs, is responsible for the flavor as well as the beneficial health effects. The extent to which these compounds are present in the final beverage depends on how the leaves are processed. To make black tea, the dried leaves are crushed, liberating enzymes that react with the catechins in the course of a few hours to produce changes in flavor and color. This is often referred to as fermentation. Green tea is not fermented; it is made by first steaming the leaves to halt any enzyme activity. Oolong tea is partially fermented. The highest concentration of catechins is, therefore, found in green tea.

The current interest in the health benefits of tea was spurred by a Dutch study published in 1993 linking the consumption of foods high in flavonoids with a lower incidence of heart disease. Especially noteworthy was the protection that seemed to be afforded by drinking four cups of tea a day and the observation that the rate of cardiovascular disease in China, where tea is extensively consumed, is one-fifth that found in the developed world. The study presented a theoretical rationale for the protective effect. High blood cholesterol is known to be a risk factor for heart disease, but we also know that the actual damage to arteries occurs when cholesterol undergoes a process known as oxidation. This process produces those highly active and potentially dangerous species we've all heard of: free radicals. The Dutch study identified catechins as potent "antioxidants," or free-radical scavengers.

Free radicals have also been implicated in the onset of cancer, so perhaps it is not surprising that lung-cancer rates in Japan are lower than those in North America, even though the Japanese smoke more. They do, after all, drink a great deal of green tea. A Chinese study showed that people who drank more green tea had a lower incidence of cancer of the esophagus, and an American study found longterm tea drinking to protect against pancreatic cancer.

Animal experiments also support the health benefits of tea. Rats fed cholesterol and green-tea extracts had lower blood cholesterol than those not given tea. Mice ingesting green tea with their drinking water developed fewer tumors when exposed to a carcinogen from smoke. Even skin cancer was less prevalent in mice exposed to ultraviolet light if those mice were given tea to drink.

Laboratory experiments confirm the antioxidant effect of tea. Free radicals chemically generated in a test tube are neutralized by tea more effectively than by the antioxidants found in fruits and vegetables. Only garlic comes close. Recently, it has become possible to measure the ability of various substances to trap free radicals in human plasma, and tea again performed in a stellar fashion. In the lab, EGCG kills cultured human cancer cells without affecting normal cells. Other studies have shown that tea increases the concentration of enzymes that remove toxins from the body and impairs the activity of another enzyme, urokinase, used by cancer cells to invade neighboring cells. But, of course, no population studies, no animal experiments, and no laboratory tests can prove that tea effectively prevents cancer in people. Only human-intervention studies can do that, and now we have the results of just such a study. Chinese researchers followed the progress of a group of people who had been diagnosed with precancerous lesions in their mouths. Normally, a high percentage of such patients would be expected to develop cancer. Half the subjects were given a mixture of green and black tea extracts to drink regularly, while the others were treated with a placebo. After six months, there was a dramatic decrease in the lesions of the people who had been drinking tea. Very impressive.

One of the results of such tea research has been the appearance of tea capsules in health-food stores. The packaging claims that one such capsule is equivalent to four cups of tea, and the product may actually deliver the goods — namely the catechins — but so far no one has carried out any intervention studies with these supplements. Even green-tea ice cream has appeared on store shelves. Just how much, and which, catechins this mouthwatering delicacy contains is an open question. Why not drink tea instead? It tastes good and may be good for us. How often can you say that?

If you're still not rushing to put the kettle on, listen to this. Tea catechins impair the activity of the bacteria in the mouth that produce cavity-causing acids, tea contains fluoride, which strengthens teeth, and there's more: you can even dye your hair with tea or soak your feet in it to reduce odor. Does tea have a downside? Children should probably consume it in limited amounts only, because it does contain caffeine — roughly a third as much as coffee — and its polyphenols interfere with iron absorption. For the rest of us, though, a "cuppa" a few times a day is just what the chemist ordered. I've taken to drinking gunpowder tea, so named because the leaves are rolled into little green balls that resemble gunpowder pellets.

The tea is green, so it has lots of EGCG. I've even thought about fortifying its antioxidant potential by brewing it up with garlic, but I'm afraid that would lead to a real tempest in the teapot.

RELISHING TOMATOES

About two thousand citizens of Salem, New Jersey, gathered in front of the county courthouse on September 26, 1820, to witness what they thought would be a suicidal act. Colonel Robert Gibbon Johnson was going to eat a basketful of tomatoes. In Continental Europe, tomatoes — or "love apples," as they were called — were believed to be an aphrodisiac and were widely consumed, but in England and America they were thought to be poisonous and responsible for a variety of diseases ranging from stomach cancer to "brain fever."

There was absolutely no evidence for this allegation other than the fact that the tomato belongs to the same botanical family as the deadly nightshade. People just assumed that it, too, must be dangerous. Green tomatoes do contain a potentially toxic compound called tomatine, but the amount of this substance in ripe tomatoes is insignificant.

Johnson was trained as a lawyer who became interested in agriculture. He had traveled extensively in Europe and had seen Italians eat plenty of tomatoes with impunity. The time was ripe, he thought, to rescue the tomato's reputation in America. In 1808, Johnson introduced the fruit to New Jersey farmers, offering a prize to the one who could grow the largest tomato, but when he announced that he would publicly eat a basketful of tomatoes, his physician predicted he would "foam and froth at the mouth and double over with appendicitis. All that oxalic acid — one dose and you're dead." Johnson's fellow New Jerseyites were inclined to believe the doctor.

When the fateful day came, Johnson, undaunted, began to eat as a local band played a funeral dirge. Men and women stared in disbelief. "The time will come when this luscious tomato, rich in nutritive value, will form the foundation of a great garden industry," the adventurous Johnson proclaimed. To the astonishment of the onlookers, he proceeded to polish off the entire basket, and his stunt was reported in newspapers across the nation. Tomato growing has since become a major enterprise in North America.

The tomato gained such popularity that importers began to supplement the local supply by shipping the fruit from the West Indies. Yes, "the *fruit*." Botanically, that is exactly what tomatoes are: by definition, a fruit is the part of a plant that surrounds the seeds and is derived from the flower's female tissue, the ovary; so tomatoes, along with green beans, eggplants, and cucumbers, are technically fruits.

In thinking of these foods as vegetables and not as fruits, we are supported by a ruling of the US Supreme Court. In the late 1800s, fruits could be imported into the United States on a duty-free basis but vegetables could not. A New York importer claimed that tomatoes from the West Indies should be duty-free since they were actually fruits. The customs agent in charge did not buy this argument and imposed a 10 percent duty. The importer appealed, and the case went all the way to the Supreme Court, which made its decision on the grounds of linguistic custom. Tomatoes are generally served with the main meal and not, like fruit, as a dessert, ruled the judges. The importer had to pay the duty, and we can legally refer to tomatoes as vegetables.

The popularity of tomatoes presented another problem. Could ripe tomatoes be shipped over long distances without softening or spoiling? Alas, the answer was no, but this obstacle triggered intense research into the chemical processes

involved in ripening. Soon it was learned that tomato plants produced a gas called ethylene, which initiated maturation. And so a plan was developed: pick the tomatoes while still green and firm, ship them to their destination, and ripen them by exposure to ethylene (which was cheap and readily available); the green tomatoes would not spoil. The plan worked. The green tomatoes were packed in cardboard boxes, shipped, and gassed. They looked great — red and appetizing — but something was missing. Taste. Biting into one of these tomatoes was not much different from biting into the cardboard boxes it came in. The ethylene gas prompted color production all right, but it did not induce the formation of the acids and sugars that characterize the wonderful flavor and smell of vine-ripened tomatoes. At this point, the genetic engineers stepped in, fueled by the hope of supplying us with tasty tomatoes year-round.

As a tomato ripens on the vine, it not only develops flavor but also begins to produce a fruit-softening enzyme called polygalacturonase. If a way were found to inhibit this enzyme, the tomato could be picked when ripe and shipped without danger of softening. Polygalacturonase is formed on the instructions from a molecule called messenger RNA, which in turn is derived from the master molecule of life: DNA. The part of the DNA molecule that holds the code for the formation of a particular RNA is what we call a gene.

Those genetic engineers had to find a way to deactivate the messenger RNA before it could trigger the formation of the softening enzyme. It was known that RNA molecules could bind to other RNA molecules if these had so-called complementary molecular structures — something like fitting the pieces of a jigsaw puzzle together.

In theory, this could be accomplished by inserting into the tomato plant an "antisense" gene that would code for the formation of another messenger RNA molecule that would bind or neutralize the "culprit" RNA. In the early 1990s, biotechnologists at a California company called Calgene managed to do exactly this, and the "Flavr Savr" tomato was born.

Optimism prevailed. In 1993, the CEO of Calgene prophesied, "We're going to sell a hell of a lot of tomatoes, and the growers, the sellers, our shareholders — everybody is going to get rich." Of course, not everyone shared this view: there was opposition from various activists concerned about the repercussions of tinkering with food-supply genetics. In public forums, they conjured up images from the cult classic movie *Attack of the Killer Tomatoes*, in which gigantic mutant tomatoes go on a murderous rampage. They threatened boycotts of companies using the genetically engineered tomatoes and promised public tramplings of the product.

This extreme opposition to the genetically altered tomato is not scientifically supportable. For awhile now, we have been altering the genetics of our food supply using various crossbreeding techniques, and there have been no calamities. There is some legitimate worry, however, when totally foreign genes are inserted into a food. For example, a gene from nuts was recently inserted into soybeans in order to increase their nutritional value as animal feed. Some worried that these altered soybeans would trigger allergies in people sensitive to nuts. Skin tests conducted with the transgenic soybeans did, in fact, confirm that this was a possibility.

In the case of tomatoes, however, no genes coding for novel proteins were introduced. The only effect of the genetic engineering was to reduce the levels of an existing compound — polygalacturonase. Even the most fervent opponents of the Flavr Savr were forced to admit that this did not represent a risk, but they still argued that genetic engineering of foods must be stopped in its tracks before potentially dangerous alterations do occur.

The bright future envisioned for the Flavr Savr has not materialized. In taste tests, panelists judged it as falling somewhere between "cardboard" tomatoes and vine-ripened ones. The major difficulty with the tomato undertaking, however, was a technical one; it seems the gene-insertion techniques employed were not quite state of the art, and the percentage of the crop that actually acquired the desired genetic trait was too low to be economically viable. Calgene insists that it is only a matter of time before this problem is solved.

So we don't have the Flavr Savr, but we do have some Israeli tomatoes that taste good even after having been shipped thousands of kilometers. Israeli researchers have developed a way to suppress the same gene that Calgene targeted using conventional crossbreeding techniques. They have successfully "mated" the tomato with the wild cherry, which has natural antipolygalacturonase activity. These succulent tomatoes are now available year-round.

And that is just how we should be eating tomatoes — year-round. At least it is if we pay attention to current nutritional research. A recent study by the Harvard School of Public

Health showed that men who eat 10 or more servings of tomato-based foods a week have a 45 percent reduction rate in prostate cancer. Spaghetti sauce is the most common tomato-based food consumed. Cooked tomatoes seem to be more protective than raw tomatoes or tomato juice, probably because heat releases the red pigment lycopene from the fruit's cells.

Lycopene, a carotenoid, may be one of the protective compounds. This has prompted the production and promotion of lycopene tablets as cancer-fighting supplements by the health-food industry. And in this case they may have something. A 1999 study conducted at the Karmanos Cancer Institute in Detroit found that prostate tumors decreased in size and that the cancer became less aggressive in men who were treated with two daily supplements of 15 milligrams of lycopene for 30 days prior to prostate surgery. Blood levels of prostate-specific antigen (PSA), a measure of tumor activity, also fell by 20 percent. To get this much lycopene from fresh tomatoes we would have to eat about a kilogram a day; of course, then we would also be ingesting other compounds, such as chlorogenic acid and coumaric acid, which have been linked to good health.

We have obviously come a long way in our appreciation of tomatoes since the days of Robert Gibbon Johnson. Still, Johnson's contribution to the pleasure of our palate (and, as it turns out, to our health) is commemorated each August in Salem, when a man in colonial garb stands on the courthouse steps and raises a fresh New Jersey tomato to his mouth as spectators yell, "No! Don't do it." Paying no heed, he takes a big, juicy bite, and then everyone else follows suit — after all, the tomato has the right chemistry.

Pour the Bubbly

A fascinating story is told by the guide on a tour of perhaps the most famous champagne house in the world. Moët et Chandon, in Reims, is the producer of Dom Perignon, the king of bubblies. The traditional saucer-shaped champagne glasses, the guide explains, were actually modeled on the shape of Madame de Pompadour's breasts. Louis xv's favorite paramour, as the fable goes, commissioned a glassblower to make the glasses in order to please the king who was so enamored of her bosom. The story goes down well with the tourists, probably even better than champagne goes down from those saucer-shaped glasses.

Whatever the real etiology of the glasses, one thing is certain: they are the wrong shape for drinking champagne. Without a doubt, the greatest appeal of this exalted beverage is the presence of the bubbles — some five million in every glass. A tremendous amount of effort goes into keeping them in the beverage. A saucer-shaped glass provides a wide surface for the liquid to contact the air, maximizing the rate at which the bubbles escape. Ideally, therefore, champagne should be sipped from a tall, narrow glass. Why should we attach so much importance to the way we consume our champagne? Because, since we probably paid a king's ransom for a bottle of "the king of wines and the wine of kings," we might as well benefit from its full, intended effect — in other words, the bubbles should burst in the mouth, not in the hand.

The solubility of carbon dioxide decreases as the temperature increases. Serving champagne cold therefore minimizes the amount of gas that escapes before we raise the glass and ensures that we experience a delightful tingling sensation when the drink comes into contact with our warm mouths. It is also important to drink champagne from a high-quality glass, one with few imperfections. Tiny air bubbles can get trapped in small nicks as the drink is poured, and the dissolved carbon dioxide then vaporizes into these bubbles. Since the carbon dioxide is less dense than the surrounding solution, the bubbles stream to the surface. For the same reason, swizzle sticks, which can have many surface blemishes, are obviously contraindicated for champagne.

So much for the bubbles. What about the drink itself? This regal beverage is produced mainly from black grapes in the Champagne region of France. From the moment the pinot-noir grapes are pressed in the vineyard, where almost fanatical care is taken to ensure that not even a trace of black skin ends up in the white juice, to the moment the cork pops, champagne receives more care and attention than any other wine in the world.

Dom Perignon, a blind monk, got the ball rolling in the eighteenth century. He discovered that if a bottle of wine was sealed tightly before fermentation was complete, the bubbles of carbon dioxide could not escape, and an effervescent drink would be produced. Due to his keen sense of smell, a result of his blindness, he was able to maximize the flavor of the wine through judicious blending of different juices. To this day, champagne is produced by the methods initiated by Dom Perignon.

The blended juices are fermented, filtered, and bottled. Pink champagne owes its tint to an added touch of red wine. Before the cork is inserted and secured with a type of wire cage, some extra sugar and yeast are mixed in to trigger the so-called secondary fermentation, which takes place in the bottle over the course of several years. During this period, a sediment consisting mostly of expired yeast cells is produced and has to be removed through an ingenious procedure. The bottles are stored with their necks tilted down in racks that can be adjusted to increase gradually the angle of the tilt. To make sure the sludge accumulates in the neck, the "remueur" walks up and down the aisles of racks giving each bottle a little twist to the right and then to the left. He can do about 30,000 bottles a day, but clearly this is not very stimulating work. The remueur has to be richly compensated, and this is reflected in the final cost of the champagne.

After the secondary fermentation is complete, the bottle is ready for the "dégorgement." The neck is dipped into a freezing brine solution until the wine and sediment in the neck solidify. In the classic process, a highly skilled "dégorgeur" uncorks the bottle, allowing the frozen plug to burst out; these days, except in the case of the real premium champagnes, machines perform the task. Sugar is then added — the amount determining whether the champagne will be brut, sec, or demi-sec — after which the bottle is quickly resealed. A few years of aging, and the cork is ready to be popped.

Contrary to North American practices, it is poor form to allow the cork to smash against the ceiling. (It is also gauche to drink champagne from a shoe.) The cork should be grasped and the bottle twisted gently while pouring so the bubbles end up in the glass, not on the floor. For those lacking in dexterity, a French inventor has devised a cork outfitted with a tab that when pulled releases the pressure. The wire cage can then be safely removed and the cork easily pulled.

Traditionalists recoil at this development, but it could help avert calamity. A couple of examples pop to mind: a few years ago, a British beauty queen was almost blinded by a flying cork; and a cork once sailed right through an expensive Victorian oil painting at the opening of an art exhibit in Bristol.

Now that we know all about bubbles and corks and secondary fermentation, just one nagging question about champagne remains. Is it true that champagne drinkers become inebriated faster? In a word, yes. Carbon dioxide accelerates the passage of alcohol into the bloodstream. The release of the gas from the champagne in the stomach causes the valve between the stomach and the small intestine to open. Since absorption from the intestine is quicker than from the stomach, the effect of the alcohol is sensed faster than it is with a nonbubbly beverage. This is especially true when champagne is consumed in an airplane, where, due to lower pressure, bubbles are even more quickly released.

And, finally, what about that old story about preserving the fizz in an unfinished bottle of champagne by suspending a silver spoon inside the neck? Be forewarned that this measure will have just the opposite of the desired effect: it will introduce nucleation sites and cause more rapid bubble loss. One last comment. Shaking a bottle of champagne before opening it is decidedly uncouth. Such an action is only acceptable if the contents are destined to be poured over someone's head after the Super Bowl, Stanley Cup, or World Series has been won. Those guys can afford to waste the bubbles.

SOLE FOOD IS A GOOD IDEA

"The sardines, Jeeves, eat the sardines!" It is with such words that Bertie Wooster, protagonist of P.G. Wodehouse's beloved stories, implores his quick-witted gentleman's gentleman to rev up his mental engine and apply it to extricating his master from yet another romantic jam. Jeeves always rises to the occasion and hatches some clever scheme to deliver young master Wooster from his predicament. Whether Jeeves actually dines on these denizens of the deep is unclear, but Wodehouse's repeated references to fish consumption and brain power attest to the prevalence of the belief in this linkage. Can eating fish really make us smarter? The answer is maybe.

The first attempt to put the long-held notion that "fish is brain food" on a scientific footing came during the 1800s, when a group of scientists discovered that the key molecule in producing cellular energy — adenosine triphosphate, or ATP — is rich in phosphorous. Since ATP provides us with the energy for thinking and is used up in the process, these scientists proclaimed that its regeneration was the key to mental acuity. And since fish is an excellent source of phosphorous, it stood to reason that it was "brain food." Today, researchers know that this is not the case. But they have discovered that another component of fish, a fat known as docosahexaenoic acid (DHA), may play a very important role in brain function.

The human brain is composed of about 60 percent fat, so in a sense we're all "fatheads." It appears, however, that it is the composition of brain tissue in terms of specific types of fats that is the key to predicting mental prowess. The earliest research suggesting such a connection focused on monkeys; when these animals were fed a diet deficient in DHA, their brains and eyes did not develop properly. This is not all that surprising, given that DHA is the primary fat found in the brain and in the retinas. Interestingly enough, supplementing the diet with DHA restored normal brain and eye development in the monkeys, demonstrating that the composition of the brain responds to dietary intake. What about humans? We're often told that we are what we eat. Do we also think with what we eat?

Some interesting evidence emerges when epidemiologists examine rates of depression around the world. It seems that the incidence is 60 times greater in some countries than in others. The United States and Canada are at the top end, while countries such as Korea and Japan have a very low incidence of depression. When fish consumption is brought into this picture, a remarkable relationship appears: countries where people consume a lot of fish have low rates of depression, and countries whose inhabitants do not consume much fish exhibit high rates. Furthermore, a study published in the *American Journal of Clinical Nutrition* has demonstrated a link between the increase in depression in North America and the decline in the consumption of DHA-rich foods. Obviously, these observations do not necessarily mean that eating fish can reduce one's depression risk, but there does appear to be some evidence to suggest this conclusion.

Low concentrations of a chemical found in the cerebrospinal fluid, 5-hydroxy-indolacetic acid (5-HIAA), have been quite conclusively linked with depression and suicide. We also know that people whose blood plasma contains low levels of DHA also have low levels of 5-HIAA. Interesting. Then consider that researchers at the University of Surrey, as well as at Purdue, have linked low blood levels of DHA to dyslexia, attention deficit disorder, and hyperactivity, and they have shown that these conditions are alleviated when a DHA supplement, now marketed as Efalex, is administered. Moreover, a study of over one thousand elderly people who were followed for nine years has shown that those with high blood levels of DHA are more than 40 percent less likely to develop dementia, including the Alzheimer's variety. Add to this the results of a Japanese study that demonstrated improved short-term memory and night vision in healthy subjects taking DHA supplements and those of a Dutch study showing that in elderly men cognitive impairment and decline was inversely associated with fish consumption, and a fairly consistent picture emerges. Healthy brain function requires adequate levels of dietary DHA.

If we are looking for even more evidence as to the importance of this particular fat in our diet, we need look no further than the first meal of our life. Breast milk is a particularly concentrated source of DHA, probably an evolutionary reflection of the importance of this fat in infant eye and brain development.

Indeed, as more and more information about the importance of DHA accumulates, infant-formula manufacturers are looking at adding it to their product. But what are fully grown people to do?

Surely, oiling our brains with DHA is a good idea. The best dietary source of this substance is undoubtedly cold-water fish like salmon, tuna, mackerel, and herring. While a couple of meals a week is thought to provide enough DHA to meet the brain's needs, not everyone is fond of fish. And some cannot partake because of allergies. Are these people destined to experience the progressive breakdown of their mental machinery? Luckily, not. There are other ways to increase levels of DHA.

The fat is found in organ meats and eggs, two foods most people have cut back on for fear of elevating their blood cholesterol; the dramatic decline in DHA intake in North America among people who don't eat fish can actually be traced to this trend. The only nonanimal source of docosahexaenoic acid is a species of marine algae, the plant that serves as the source for the DHA in the flesh of fish. This is an impractical source for humans, but techniques have been developed to extract the chemical and formulate it into supplements. Fish-oil supplements are also available and are an alternative to eating fish, but they don't solve the allergy problem, plus there's the question of the smell — one whiff is enough to make a host of cats drool.

If supplements don't appeal, there is another solution. Our bodies can manufacture some DHA if fed the right raw material — namely, the essential fat alpha-linolenic acid (ALA). This stuff is found in soybeans, canola, nuts that grow on trees, and, particularly, flaxseed. Cooking with canola oil, making salad dressing with flaxseed oil, and snacking on nuts should keep the brain well lubricated. How about a salad with pine nuts and a flaxseed-oil dressing? If even this doesn't sound attractive, then you'll just have to focus on research being carried out at Guelph University, where scientists are feeding fish meal to cattle in an attempt to increase the DHA content of the meat. Apparently this works, and allergies to the fishy meat have not been noted.

All of this, I'll admit, sounds a little complicated. Maybe it's because I find some aspects of the research quite perplexing and difficult to interpret. But I do have an excuse. You see, my brain may not be well oiled — because of an allergy, I haven't eaten any fish since I was a young boy.

SOUPY SCIENCE

When I feel a cold coming on, I like to approach the problem scientifically. This means dissolving some cysteine, some alliin, some piperine, and some potassium ions in hot water and drinking the concoction. I cannot claim any originality here. Way back in the twelfth century, Moses Maimonides, the noted physician and philosopher, recommended much the same mixture of chemicals as a treatment for the common cold, only he called it chicken soup. It proved to be a lot more popular than kissing the hairy muzzle of a mule, a therapy advocated by Pliny, the ancient Roman philosopher. It was also more acceptable to Europeans than soup made from freshly killed snakes, the preferred Chinese remedy.

Can this mixture of chicken meat, vegetables, and spices actually have therapeutic value? Researchers at the University of Nebraska Medical Center in Omaha showed that chicken soup does inhibit the movement of white blood cells called neutrophils. These cells go to the site of infection and release enzymes that attack bacteria and viruses, but they also attack the body's own cells, causing inflammation. A sore, inflamed throat is a typical cold symptom. Somehow, chicken soup reduces the inflammation effect without reducing the antiviral activity.

Cysteine, a key amino acid in the development of meat flavor, is plentiful in chicken flesh and makes a valuable contribution to taste — but it may do more than that. A close chemical relative of cysteine, N-acetylcysteine, is a medication often used to thin mucus in the bronchial tract, making it easier to expel. There is good reason to believe that cysteine can have a similar effect in the body, leading to decongestion and easier virus elimination.

Some support for this possibility comes from a rather imaginative study carried out in the late 1970s at Mount Sinai Hospital in Miami. Researchers fitted the noses of 15 volunteers with miniature sensors capable of measuring the speed with which mucus is eliminated. They were then asked to drink either chicken soup, hot water, or cold water. Chicken soup was far more effective than cold water and was about 10 percent better than hot water at speeding up mucus flow. Not such impressive results when you consider that the effect lasted only half an hour. Now, I like chicken soup, but every half-hour?

Mucus thinning, though, may not be the most important role for cysteine. This compound is one of the body's precursors for glutathione, a very important component of our immune system. An increase in glutathione levels could certainly be expected to boost the immune system. Maybe the reason the Mount Sinai researchers did not find a significant improvement in mucus flow after substituting chicken soup for hot water was that they were using a bland soup. Just spice up that soup and then watch the mucus flow. Haven't we all experienced the effects of garlic, black pepper, or chilies at one time or another? These substances almost immediately trigger rapid nasal secretions — "salsa sniffles" are legendary.

Garlic contains alliin, pepper has piperine, and chilies are loaded with capsaicin. Alliin, as it turns out, bears a strong chemical resemblance to the active ingredient in the medication Mucodyne, which is used to thin lung secretions. William Harvey, the seventeenth-century physician who developed the theory of blood circulation, recommended to his patients that they place a clove of garlic in the toe of their shoe to treat congestion. Scientists have actually confirmed that if this practice is followed, the scent of garlic is detectable in the breath, confirming absorption into the bloodstream, but I think it's better to put the garlic into the soup.

Black pepper and chilies both have compounds that resemble guaifenesin, an expectorant found in over-the-counter cold remedies. So does ginger, which some adventurous cooks add to chicken soup. The Japanese have their own ginger-based remedy for the common cold: they mix grated ginger and sugar with hot water or hot sake (I'm sure that after they've dosed themselves with this concoction they forget all about their colds). This raises an interesting point.

Could it be that the real secret is the heat and not the dissolved chemicals? This question was addressed in the late 1980s by the French Nobel laureate Andre Lwoff and his colleague Aharon Yerushalmi at the Weizmann Institute in Israel. Working with the assumption that the cold virus prefers to live in the nose than anywhere else in the body because the temperature there is about four degrees lower, these researchers developed a high-tech virus annihilator. This instrument, known as the Rhinotherm, delivered a spray of hot, humid pressurized air via a tube held about an inch beneath the nose.

Looking at the early results, the researchers claimed that after undergoing three half-hour sessions a few hours apart, three quarters of the test subjects recovered from their colds within a day. Unfortunately, follow-up studies conducted by others did not support this early work. A very well-designed study done at the Cleveland Clinic actually showed that patients fared worse with the Rhinotherm, suffering more sniffling and greater congestion. Perhaps someone should investigate loading the device with chicken soup instead of water.

So we're back to chicken soup. Maybe it doesn't cure the common cold, but it can certainly help adjust our electrolyte balance. The flow of fluids into and out of cells is regulated by the presence in the body of the minerals sodium and potassium. Dehydration, a symptom of a number of illnesses, can upset this balance and lead to yet other symptoms ranging from tingling sensations to diarrhea. Vegetables, parsnips in particular, have a very high ratio of potassium to sodium and can correct electrolyte imbalances. In fact, when Barney Clark, the first American ever to have a heart transplant, suffered seizures due to a mineral imbalance caused by taking various medications, the prescribed treatment was chicken soup — delivered directly into the stomach by means of a feeding tube.

Let's now put all of this information together. In a big pot, place two liters of cold water. Add six sliced carrots, three parsnips cut in half, a whole onion, a cubed celery root, four celery stalks, half a green pepper, and eight peeled cloves of garlic. Next comes my secret ingredient. I use Knorr chicken-bouillon cubes — six of them. Why should I bother to extract the cysteine from the chicken when someone else has already done so more effectively? Now, add black pepper to taste and some fresh dill. Simmer for 40 minutes and then consume at least once a day. Your cold will disappear within a week, guaranteed. Difficult cases may take a full seven days.

SOYBEANS, CABBAGES, AND BREAST CANCER

Why all the excitement about soybeans? Because Japanese women have one-quarter the breast-cancer rate of North American women, and Japanese women eat a lot of soy products. This does not necessarily mean that soy consumption has anything to do with breast cancer (after all, there is a very strong association between the disease and the wearing of skirts, and obviously nobody thinks that the wearing of skirts causes breast cancer). Still, when we start considering the scientific evidence, it begins to look as if the soy connection is more than just a chance association.

Our story begins in the 1940s, when Australian farmers noticed that sheep grazing on a certain type of clover failed to reproduce normally. The urine of these sheep was found to contain a high level of a compound called equol, which had previously been found in the urine of pregnant horses. Bacteria in the sheep's intestine, as it turned out, had converted a naturally occurring compound in clover to equol, and equol was known to have biological activity similar to estrogen. It came as no great surprise to scientists that an estrogen-like substance should interfere with fertility, since estrogen was already known to play an important role in human reproduction. They began to wonder if other foods had naturally occurring compounds that possessed estrogenic activity.

Enter the soybean. Researchers discovered that this Asian staple contained compounds, collectively known as isoflavones, that did indeed exhibit estrogen-like behavior. Genistein and daidzen, in particular, were of interest because they were partially excreted in the urine and could be correlated with the amount of soy in the diet. This raised eyebrows, because it had already become apparent to scientists that estrogen and breast cancer were somehow connected. Women who are exposed to more estrogen over a lifetime were known to have a higher risk of contracting the disease, including women who come into puberty early, reach menopause late, or have few or no children. In other words, it appears that any factor that lowers the total number of menstrual cycles over a lifetime lowers the risk.

Now let's return to Japanese women. They have longer menstrual cycles, averaging 32 days compared with the North American rate of 29 days. This could mean 30 to 40 fewer periods in a lifetime. They also have up to a thousand times more phytoestrogens in their urine than do North American women. But the soybean plot really thickens when we note that Japanese people consume 30 times more soy products than we do and that those who migrate to North America and take up the North American diet and lifestyle show cancer rates comparable to our own.

Recent research has even revealed a possible mechanism for the connection between isoflavones and breast cancer. Some cells in breast tissue are known as estrogen responsive, meaning that they contain certain proteins (estrogen receptors) to which estrogen can bind, very much in the fashion of a key fitting into a lock. This binding unleashes a sequence of events in the nucleus of the cell, eventually leading to the manufacture of certain proteins that trigger cell proliferation. Such abnormal cell multiplication can lead to cancer. Isoflavones, it seems, are actually "weak estrogens." They fit into estrogen receptors but do not stimulate any cellular activity. At the same time, they prevent estrogen from binding with the receptor. It is as if the wrong key had been inserted into the lock: the key cannot be turned, but it effectively prevents another key from being inserted.

So much for associations and theory. What practical evidence can we muster to show that soy consumption may actually prevent breast cancer? A number of animal studies have demonstrated that the consumption of soy or isolated isoflavones reduces tumor development. Human data is less direct, but researchers have compared groups of breast-cancer victims with matched controls and noted a decreased risk of up to 50 percent in premenopausal women who consumed soy daily.

A classic study conducted in Singapore showed that breast cancer rates correlated inversely with the amount of soy protein eaten on a regular basis. More than 20 studies of Asian women have shown that even one cup of soy milk or half a cup of tofu a day is associated with reduced cancer risk. It has also been discovered that menopausal women who start eating 20 grams of soy protein powder daily (roughly equivalent to a soy burger, a cup of soy milk, or a serving of tofu) show a reduction in the severity of menopausal symptoms, and an added benefit is increased bone density in the spine. As far as premenopausal women go, the same kind of diet lengthens their menstrual cycles by 2.5 days and dramatically increases the isoflavone content of their urine. It is clear that soy has estrogen-like activity.

And now we find out that genistein, the main isoflavone, may have yet another effect: it appears to decrease the growth rate of the blood vessels that nourish tumors. This inhibition of "angiogenesis" may turn out to be the most important anticancer effect, and it may even explain why men whose urine has high levels of genistein seem to be protected from prostate cancer. Although it may be that isoflavones are the most interesting anticancer compounds in soybeans, there are others — folic acid, for one, has been shown to prevent mutations in DNA. There seems to be no end to soy's benefits. Soy protein can even lower cholesterol, and when it is mixed in with ground beef it reduces the amount of carcinogens that form during the broiling process.

There are, however, some inconsistencies in the soy saga. One study of Japanese women showed that those with breast cancer had consumed no less soy than a control group unaffected by the disease. Chinese women, who eat only about a third of the soy-based foods that the Japanese eat, have the same low rate of breast cancer. Of course, it is possible that a certain amount of soy is protective but eating more carries no further benefit.

Even though there are uncertainties about the role of soy in breast-cancer prevention, there is certainly no harm in increasing one's intake of isoflavones. Remember, though, that not all soy products are equally good sources: soy oil, tofu hot dogs, and tofu ice cream are poor sources, but tofu itself, soy milk, tempeh, miso, soy flour, and textured soy protein contain good doses of isoflavones. But before we get too carried away with the isoflavones, it is important to realize that breast cancer is a complex disease with many possible contributing factors. The disease is age related and linked to excessive alcohol consumption. There may be a connection to high levels of certain fat-soluble pesticides. As far as the fat content of the diet goes, studies have been ambiguous; some noted an increased risk with saturated fat, while others showed high risk with increased carbohydrate consumption. Monounsaturated fats like canola oil or olive oil appear to be the best choices. Exercise, fruits, and vegetables are protective.

Particularly effective are the cruciferous vegetables, like broccoli and cabbage. These contain indole-3-carbinol, which protects against estrogen-stimulated breast cancer. Could this be the reason why, prior to unification, the breast cancer rate in East Germany, where inexpensive cabbage was a dietary fixture, was much lower than in more affluent West Germany? Cabbage is easy enough to prepare, but what do you do with soybeans? You can soak them in water overnight and then cook them like any other bean; alternatively, you can just roast them in the oven or in the microwave and eat them as a snack. And then there's tofu. How about an anticancer combo of cabbage and soybeans? I'm working on it, and the taste is fine — the ill wind that is produced may well blow good.

THE FEET OF GOD

The French poet Léon-Paul Fargue inhaled deeply as he beheld the Camembert cheese: "Ah, the feet of God!" he exclaimed. A curious, yet accurate, description of the mix of butyric acid and methyl mercaptan, the compounds that characterize the aroma of this cheese. The same compounds are actually found in the daunting scent of sweaty feet.

Cheeses such as Camembert, Brie, Roquefort, and Limburger are ripened through treatment with molds or bacteria. These microorganisms produce a range of enzymes that slowly degrade the fats and proteins in the cheese to produce a variety of flavorful, albeit smelly, compounds. At the same time, these chemical changes also cause the cheese to soften. That's why surface-ripened cheeses such as those I have mentioned are always thin wheels; if they weren't, the outer layers would liquefy while the inner core remained hard. In case you're wondering, the proper time to eat one of these cheeses is when the enzyme activity has just reached the center, making it a little runny — or "au coulant," as a proper turophile (cheese lover) would say.

The microbes that release butyric acid from fats and methyl mercaptan from proteins in cheese are very similar to the organisms that lurk between our toes. The French expression "ça scent du Roquefort, Jacques" is therefore a most appropriate means of alerting a victim of *brevibacterium epidermis* to

the existence of a problem. Experiments have even demonstrated that extracts of toenail clippings and Limburger cheese have a very similar fatty-acid composition. Perhaps even more interesting is the observation made by researchers, using naked volunteers as mosquito bait, that the type of mosquito which dines exclusively on the ankles and the feet is also attracted to Limburger.

Surface-ripened cheeses can attract more than mosquitoes — bees, for example. One of the compounds formed by microbial activity in ripening cheeses, 2-heptanone, also, by some quirk of nature, happens to be the chemical that bees secrete to warn others of impending danger. Therefore, it's probably not a good idea to eat Roquefort cheese near a beehive.

Indulging in a little Roquefort, or Brie, or Camembert anywhere else, though, is a fine idea. Their microbial by-products make for a dynamic, exhilarating flavor. But have you ever wondered how we ever managed to discover what inoculating a cheese with a slimy mold could do? By accident, of course. Cheese has been with us a very long time. According to popular lore, sometime around 2300 B.C. a nomad in the Middle East noted that the milk he was carrying in a bag fashioned from the stomach of an animal curdled. Today we understand why: an enzyme called rennet, which is found in the stomach lining of young animals, causes the proteins that are suspended in milk to coagulate into curds and separate from the liquidy whey. Remember Little Miss Muffet eating her "curds and whey" and her unfortunate encounter with a spider?

Soon, people began to make cheese on purpose because it would keep longer than milk. They discovered that virtually any type of milk would do — yak, reindeer, buffalo, and guinea-pig milk could all be curdled. Camel milk, however, was the exception; its protein composition is different from that of other milks and is unaffected by rennet. Only recently have scientists discovered that by adding calcium phosphate and about 10 percent sheep milk we can curdle camel milk. Camel-bert here we come!

As cheese making became popular, people began to manufacture large supplies and store them for future use. Caves were ideal storage places because their cool temperatures slowed spoilage. In some instances, however, airborne mold spores would find a particular cheese to be an ideal breeding ground and then rapidly cover it with a fuzzy layer. Some adventurous soul tasted that fuzzy stuff and discovered that it was delicious: the surface-ripened cheese industry was thus established. Roquefort is a good example. The limestone caverns at Roquefort in France harbor spores of *penicillum roquefortii*, and these spores are what gave rise to the classic cheese. These days, a suspension of the mold is sprayed over the cheese, which is then pierced with stainless-steel needles to allow the mold to penetrate thoroughly.

At one time, copper needles were used in this process, leading people to believe that the pigment found in blue cheeses was the product of dissolving copper. That color is really produced by the mold and has nothing to do with the needles. The ewe's milk employed in the production of Roquefort cheese is sometimes bleached to make the blue pigment stand out even more.

Camembert, Brie, and Limburger were all developed in a fashion similar to Roquefort. In each case, the particular flavors developed when microbes unique to the area colonized the cheese. Because these cultures have become commercially available, we can now buy, for example, Canadian Brie. Cheese connoisseurs argue that while this may be good, it is not "real Brie" — the authentic stuff has to come from French cows that feed on French grass.

And now to the crux of the moldy cheese problem — the battle between the rindophiles and the rindophobes. Do we, or do we not, eat the rind of the cheese? Cheese aficionados generally follow the wisdom of the great Charlemagne. The Holy Roman Emperor, as the story goes, once stopped in at the residence of one of his bishops. It being Friday, he was offered a meal of cheese instead of meat. Having never seen moldy cheese before, Charlemagne proceeded to cut off the rind and eat the inside. "Why do you do that, Lord Emperor?" the bishop asked. "You are throwing away the best part!"

Charlemagne sampled the rind, and he liked it so much that he asked the bishop to send him two cartloads of the cheese every year. He was right — the outside of the cheese is remarkably flavorful — but here in North America we tend to be very finicky when it comes to putting fuzzy molds into our mouths. We worry that these substances may somehow undermine our health, and sometimes we express that worry in an absurd fashion.

Recently, I was preparing a lecture on cheese and planned to cap it off with a tasting session, so I went out and purchased a variety of cheeses ranging from an unpretentious cottage cheese to an aggressive goat cheese. I also wanted to include some noble, elegant Brie and planned to serve a local Brie of questionable breeding for the purposes of comparison, but I found something even better to create the contrast: believe it or not, an American Brie, in a package, with the rind already removed. The ultimate convenience food for a rindophobe. I had to try it. I sniffed it first. No hint of the "feet of God" — instead, something more akin to the armpit of plastic. Nevertheless, I tasted the stuff. Searching for an analogy, melted floor wax came to my mind.

I think I'll stick to the real stuff until something better comes along. Actually, I just heard about a Japanese blue cheese made with soy flour and ripened with a special Oriental fungal mold. I can hardly wait to taste that one.

The Food of the Gods

Let's get something straight right off the bat. Chocolate is not an aphrodisiac, and it does not cause people to fall in love. It may, however, lift our spirits and perhaps even offer some protection from the damaging effects of high blood cholesterol. The aphrodisiac story is an ancient one. It goes all the way back to 1519 and the first visit of the Spanish explorer Hernando Cortés to Mexico. Cortés found much to his liking there, in particular the Aztec princess Doña Marina. His affection was evidently returned, because the princess introduced Cortés to a drink made from the pods of a tree the Aztecs called "chocolatl," or "food of the gods." The concoction was also laced with dried chili peppers and, as Doña Marina said, could "stimulate amorous adventures."

Cortés must have been impressed by the effects, because on his return to Spain he presented Emperor Charles V with a gift of cocoa, as we call the substance today. Within a few years, people throughout Europe were indulging in chocolate and singing its praises. Except nuns, that is: they were forbidden to partake because of the potential consequences. That prohibition was unnecessary, because, alas, chocolate does not have aphrodisiac properties. The myth can be ascribed to the presence in chocolate of general stimulants, like caffeine, theobromine, and the newly discovered anandamide.

Chocolate contains over three hundred compounds with imposing names such as furfuryl alcohol, dimethyl sulfide, phenylacetic acid, and phenylethylamine. It is this last, amphetamine-like substance that has been alluringly labeled "the chemical of love." People in love may have higher levels of phenylethylamine (usually abbreviated as PEA) in their brains — this surmise is based on the fact that their urine is richer in a metabolite of PEA. In other words, people who are thrashing around in the throes of love pee differently from those who aren't.

This observation has stimulated the following thought process: falling in love is associated with higher PEA levels; chocolate contains PEA; therefore, chocolate can make us fall in love. Not so. A person's blood levels of phenylethylamine do not rise after he or she eats chocolate. Most of this enchanting compound is metabolized during digestion. Furthermore, chocolate isn't even a very good source of PEA — sauerkraut is far better, but that doesn't make for nearly as good a story on Valentine's Day.

Why are we so infatuated with chocolate? Could it have something to do with anandamide, a compound the brain normally produces to signal pleasure? Indeed, anandamide receptors can be stimulated by foreign substances such as tetrahydrocannabinol, or THC, the active ingredient in marijuana. It bears a chemical similarity to anandamide and therefore triggers pleasurable sensations. Chocolate contains anandamide itself, so wouldn't it have the same effect? Probably not.

The amount of anandamide in chocolate is actually very small when compared with the amount produced naturally by the body: an adult would have to eat more than 10 kilograms of chocolate to get a buzz (well, maybe a little less). A couple of other recently isolated compounds from chocolate, N-oleylethanolamine and N-linoleoylethanolamine, inhibit the breakdown of anandamide and may result in higher blood levels of the substance, but rest assured that chocolate does not cause a cannabis-like high.

There is yet another candidate for the secret ingredient responsible for the appeal of chocolate. Endorphins are a class of naturally occurring substances synthesized in the human brain in response to a variety of stimuli. In general, they have been linked to effects similar to those caused by opium. "Runner's high," for example, has been ascribed to endorphin production. According to some researchers, chocolate stimulates endorphin release. This hypothesis is based on the observation that when volunteers are treated with naloxone, a drug that blocks the effect of endorphins, they get no more pleasure from eating Snickers or Oreos than from eating celery sticks.

Chocolate is, of course, also high in carbohydrates — mostly sugar. Numerous studies have shown that carbohydrates increase the levels of an important brain chemical known as serotonin, which has decided antidepressant effects; in fact, several common antidepressant medications work by increasing concentrations of serotonin in the brain. But do we really have to comprehend complex brain chemistry in order to explain our love affair with chocolate? Can it not be that this combination of flavors, sugar, and fats — which melts exactly at body temperature — just tastes great? Sure it can. This, however, raises another problem. Something that tastes so good can't possibly be good for us.

Recent research mercifully suggests that chocolate may actually have some redeeming nutritional features. Although it is high in fat, the specific types of fat it contains do not seem to raise cholesterol. Then there is the presence of polyphenols. These are the same compounds that have received a great deal of publicity in connection with the supposed benefits of red wine. Laboratory studies have shown that they can prevent the oxidation of LDL cholesterol (the "bad" cholesterol) to a form that damages arteries. A typical chocolate bar actually has the same phenolic content as a glass of red wine; the darker the chocolate, the more phenolics it contains.

While no study has shown a reduction in heart disease related to chocolate consumption, a provocative study involving human volunteers has shown that 35 grams of defatted cocoa (about the amount found in seven cups of hot chocolate) has a significant impact on preventing LDL oxidation.

Although the polyphenol evidence may not be enough to exonerate chocolate of nutritional crimes, everyone agrees that smelling chocolate is harmless enough. In fact, it may be beneficial. A study conducted at Yale University has shown that students exposed to chocolate smell while studying for an exam can recall the material better if they are also exposed to chocolate smell while writing the exam.

Even more stimulating is the research of a Chicago neurologist, Alan Hirsch, who fitted the penises of volunteers with little blood-pressure cuffs and determined that certain smells, chocolate among them, increased the pressure. While the impact of this study is unclear, why not take a good whiff of those dark chocolates before passing them around on Valentine's Day?

Granted, our scientific meanderings have not uncovered a consistent explanation for the appeal of chocolate, but the appeal is clearly there. In fact, a recent Gallup poll showed that a majority of British women would be willing to give up sex for chocolate. I've got to try that British chocolate.

THIS PULP ISN'T FICTION

It is time to come to the rescue of the much maligned albedo, the stringy stuff found on the inner skin of citrus fruits. And while we're at it, let's put in a good word for orange-juice pulp. I've often watched people do battle with the albedo, struggling to remove every last vestige before popping a segment of naked orange into their mouths — these are probably the same people who choose filtered orange juice over the pulpy variety. Too bad, because both the albedo and the pulp are good sources of pectin, a type of fiber that shows great nutritional promise.

Most people associate pectin with jams and jellies, not with cleaning out arteries, controlling blood sugar, or preventing cancer. But it just may do all of these. Pectin is a kind of carbohydrate "glue" that helps hold plant cells together. It can be extracted with boiling water from apple cores or from citrus albedo and then processed into a powder for thickening jams. In the presence of appropriate amounts of acid and sugar, the long pectin molecules bind to each other to form a three-dimensional lattice that traps water molecules. Some fruits, such as grapes and most berries, have enough pectin to produce jams on their own, but apricots and strawberries, for example, need added pectin if the jam is to achieve the proper consistency.

It is precisely this thickening ability that may be responsible for some of pectin's reported health benefits. Pectin is a form of fiber — the human digestive tract cannot break it down so that it may be absorbed into the bloodstream. It therefore remains in the small intestine, where it forms a gel just like that found in jam. This gel traps and eventually eliminates some of the bile acids that are secreted by the liver to aid fat digestion. The liver then has to make more bile acids to replace the lost ones; since the raw material used for bile-acid synthesis is cholesterol, the end result is a lowering of blood cholesterol.

This is not just a theoretical hypothesis. Researchers at the University of Florida have already shown that citrus pectin can keep pig arteries free of plaque and may actually clear blocked arteries. We cannot be sure that the same effect will be seen in humans, but it is likely. A number of studies have demonstrated that pectin can lower cholesterol by anywhere from 5 to 19 percent. This is in the same range as some cholesterol-lowering medications.

But how much pectin do we require to lower cholesterol appreciably? And in what form does it need to be? The consensus of the studies is that about 15 grams are required if powdered pectin is used but less if it is present as a food component. When we consume pectin extracted from apples, for example, the effect on our blood cholesterol is not as great as it is if we eat the apples themselves. In one study, the addition to the diet of two or three apples a day, each containing two to three grams of pectin, lowered the "bad" cholesterol in the blood by some 11 percent, but when this amount of pectin was ingested in extract form, the results were far less impressive.

A likely explanation is that the apples also contain vitamin C, a vitamin that helps pectin reduce blood cholesterol since it is needed by the body to convert cholesterol into bile acids. This is another good example of how a whole food is more than just the sum of its parts and how supplements may not be the best way to consume nutrients.

There are many good food sources of pectin. Grapefruit is an excellent one, especially if the albedo and the membranes that divide the segments are eaten. A couple of grapefruits a day can have a measurable cholesterol-lowering effect — so can oranges, carrots, and soybeans. Pectin can even regulate blood sugar, so foods containing pectin are good for diabetics.

Again, the effect is likely due to the enhanced viscosity of the intestinal contents due to the gelling ability of pectin: sugars in the intestine diffuse more slowly through the thickened material and are therefore absorbed at a slower rate. Pectin may also slow the rate at which the stomach empties, again retarding the absorption of sugars from the intestine.

And if all this weren't enough to send us scurrying for an extra dose of albedo, just mull over what researchers are discovering about the connection between pectin intake and the most dreaded of all diseases: cancer. Since pectin is indigestible, it eventually makes its way to the colon. Here it serves as food for the many bacterial species that inhabit our dark innards. We may not be able to digest pectin, but these creatures certainly can. Some of the products of this bacterial fermentation are compounds referred to as short-chain fatty acids. These acids — butyric and propionic, for example — have been studied in tissue culture and are shown to be protective against colon tumors.

Even more exciting are the studies that have been carried out using something called "modified pectin." Treatment of pectin with an alkali and then an acid produces soluble fragments that can be fed to test animals. Early results indicate that this type of pectin may prevent cancer cells from attacking normal cells and may therefore reduce the spread of the disease. More work needs to be done, but a cancer-fighting pill based on pectin is certainly within the realm of possibility.

In the meantime, it is a good idea to increase your intake of foods that contain pectin. Go for the juice with the pulp and eat a couple of apples and grapefruits every day. And stop struggling with the albedo clinging to your orange or your banana. Remember that the stringy stuff just may have the right chemistry.

VEGETABLE À LA ALA

I'd been wanting to taste some purslane for a long time, and my chance finally came when a couple of friends who were going to vacation in Crete agreed to bring me back some of this leafy vegetable, which the locals call "glistridia." The reason for my interest is that purslane has been linked with the rather remarkable health of Cretans — they apparently have the lowest rate of heart disease in the world.

This story really starts just after World War II, when Ancel Keys, an American epidemiologist, examined worldwide heart-disease statistics and noted that there were dramatic geographic differences. Keys then embarked on a more formal undertaking, which he called the Seven Countries Study, examining the factors that could account for these differences. The focus was on Finland, the United States, Holland, Italy, Yugoslavia, Northern Greece, and Crete, and the evidence gathered clearly indicated that Finland had the highest incidence of heart disease and Crete the lowest. Furthermore, the Cretans also seemed to have a low cancer rate and a particularly long life span.

An examination of diet was an obvious starting point in the quest to explain these remarkable differences. The Finnish diet is very high in fat, particularly saturated fats from meat and dairy products. Since scientists had already suggested a link between a rich diet and heart disease, the high incidence of cardiovascular disease among the Finns was hardly surprising; what struck the researchers as astonishing, though, was that the study showed the Cretans were eating just as much fat. But this fat was of a different kind: it was olive oil.

Olives and olive oil have been Cretan dietary staples for over three thousand years. We know this because in 1960 a Greek archeologist made an amazing discovery. In a deep well, he found a bowl of olives that dated back to about 1500 B.C. A violent earthquake is known to have struck Crete at that time, and the olives had probably been lowered into the well to placate the gods of the underworld who were so violently shaking the earth. Could the long-established practice of eating olives and olive oil account for the earthshaking results of the Seven Countries Study?

Olive oil belongs to the monounsaturated category of fats. These do not raise blood cholesterol the way that saturated fats do, and indeed some scientists have ascribed the health benefits of the Mediterranean diet to olive oil. But this may be simplistic: the Japanese island of Kohama has a heart-disease rate that rivals that of Crete, and there isn't an olive in sight. The diet on Kohama is based on copious amounts of canola oil and soybean oil, which are chemically quite different from olive oil. They fall into the category of polyunsaturated fats, which, like the monounsaturates, are more friendly to our blood-cholesterol profile than the saturated fats. Still, other polyunsaturates, such as sunflower oil, have not been linked with such impressive health benefits, so what is it about canola and soybean oil? They're high in a particular type of polyunsaturated fat, known as alpha-linolenic acid (ALA), and therein may lie at least part of the solution to the Cretan and Japanese mystery.

High blood levels of ALA have been associated with a lower incidence of heart disease and stroke. A study done recently at the University of California actually found that the incidence of stroke dropped by 37 percent for every 0.13 percent increase in ALA in the blood, probably due to the reduced risk of blood-clot formation. This may then explain the health of the Kohama natives, but what's going on in Crete? Olive oil doesn't have any appreciable amount of alpha-linolenic acid. Ah, but what the Cretans pour their olive oil *on* does: you guessed it, purslane.

This leafy green vegetable is an outstanding source of ALA; so are walnuts, which are liberally consumed in Crete. Still, linking these foods with good health in the absence of further evidence would not be good science. After all, Cretans also eat loads of snails, and no one has suggested that this is the cause of their good health. Obviously, the alpha-linolenic-acid argument needs more support.

Serge Renaud, a French epidemiologist, decided to put the Cretan diet to the ultimate test. He wanted to determine whether it could actually prevent heart attacks in patients at risk. In a classic study, he enrolled 605 volunteers who had already suffered heart attacks; half would follow the low-fat diet recommended by the American Heart Association and half would eat the Cretan way — this meant lots of bread, green vegetables, nuts, fruit, wine with meals, and, of course, olive oil. Test subjects ate fish frequently but consumed only very small quantities of meat and dairy products. They were ultimately followed for a longer period, but Renaud brought the study to a dramatic conclusion after just four years. The low-fat group was already showing a heart-attack rate six times greater than the group eating the Cretan way, and Renaud's ethics dictated that this information be revealed so that those wishing to alter their diets could do so.

Throughout the study, the blood cholesterol, triglycerides, and blood pressure of test subjects were closely monitored. Surprisingly, in light of the sensational results, these measurements were virtually identical in the two groups, but there was one interesting difference: the subjects following the Cretan diet had 70 percent more alpha-linolenic acid in their blood. It seems the ALA plot thickens as the blood thins.

The food industry has taken note and is in the process of designing foods with increased levels of ALA. Would you believe eggs? Flaxseed is especially rich in ALA, and chickens will readily eat it. The eggs they lay have almost 20 times as much of this fatty acid as regular eggs. Only time and more research will tell whether we will eventually see an advertising slogan along the lines of, "Lower your risk of heart disease: eat eggs!"

We don't really have to depend on the dietary regimen of chickens to benefit from the alpha-linolenic-acid content of flaxseed. We can use flaxseed oil in salad dressings and perhaps even sprinkle the seeds themselves onto the salad. And flaxseed bread tastes surprisingly good. Now, just imagine how super-healthy those Cretans would be if they dressed their purslane with flaxseed oil and ate it with flaxseed bread.

So we're back to the purslane. My friends did bring me some from Crete; while the flavor is interesting, it takes some getting used to. However, a Cretan dessert I sampled made with ground nuts, sesame, cinnamon, and grape juice ("moustalevria") was an instant hit, and, apparently, it is widely consumed on the island. Maybe all those antioxidants in the grape juice also play a role in the legendary Cretan longevity.

The recipe for this delicacy came from a tourist magazine that also featured an ad for Creta Farm, a hog-breeding enterprise founded in 1970. The advertisement proudly noted that Creta Farm had launched a cold-cuts department, which was now doing a brisk business. I can't help but think that these processed meats will squeeze some of the fish, the fruits, and, of course, the purslane, out of the Cretan diet. Will the next generation of Cretans enjoy the same remarkably low heart-disease rates as the current one?

CHEMICAL CRIMES

A LOVE POTION MOST LETHAL

The British tabloids couldn't get enough of the story. "Love Drug Kills Typists," screamed the *Daily Mirror* headline on April 29, 1954, introducing the public to one of history's most bizarre poisoning cases. By the time Arthur Ford was convicted of manslaughter a couple of months later, readers who had devoured every word written about the trial had become educated in several aspects of chemistry and toxicology.

Ford was a middle-aged man, who, after the war, had found a job as office manager with a London pharmaceutical company. The married father of two children became extremely fond of one of his secretaries. She rebuffed his overtures, probably just fanning his passion. Then, as Ford saw it, a solution to his romantic problem appeared. A customer came in inquiring about a drug called cantharidin, which he understood could be used to remove warts. The mention of the term "cantharidin" triggered a flood of memories in the rejected lover's mind. He recalled that while he was serving in the army his fellow soldiers had talked about using Spanish fly as an aphrodisiac to stimulate their reluctant partners. Cantharidin, he remembered, was the supposed active ingredient.

Ford immediately asked his firm's senior chemist whether they did indeed stock cantharidin. When asked why he was interested, Ford mumbled something about one of his neighbors breeding rabbits and possibly needing some cantharidin to facilitate the process. The chemist, however, emphatically pointed out that cantharidin was very dangerous and even small doses could prove lethal. Upon hearing this, Ford seemingly abandoned his interest in rabbit breeding.

But the allure of easy sexual conquest proved too much for him. Arthur Ford stole a small amount of cantharidin from the stock bottle, introduced it into a chocolate-covered coconut-ice-cream treat he had purchased, and shared it with the object of his desires. Another secretary asked for a taste, and within a short time all three were in hospital complaining of excruciating stomach pains and headaches. The two women died the next day, but the shaken would-be Casanova recovered.

The postmortem revealed the presence of cantharidin in both bodies, and a distraught Arthur Ford could not contain his guilt. He told authorities how the office romance that never was resulted in the accidental death of two innocent people. The judge maintained that Ford had been warned about the dangers of cantharidin by an expert and therefore sentenced him to five years for manslaughter.

Curiously, the same year, the *British Medical Journal* reported on another cantharidin-poisoning case that was just as bizarre. A fisherman who thought he could catch more fish if his bait was more sexy placed the bait, some cantharidin, and some water into a bottle and shook the mixture with his thumb clamped over the bottle's mouth. While he was baiting the hook, he pricked his thumb and, according to common practice, began to suck it. Within six hours he was dead. Cantharidin is not particularly water soluble, and some of the undissolved substance had apparently stuck to the unfortunate man's thumb.

These tragedies would not have occurred if cantharidin did not have an undeserved reputation as an aphrodisiac. The myth is based on cantharidin's ability to affect genital erectile tissue in both men and women. What it actually does — and this can hardly be described as pleasurable — is irritate the urogenital tract. Cantharidin's reputation does not stem from a history of successful use; it is built upon the supposed antics of Casanova, the legendary Spanish lover. This eighteenth-century playboy's bedroom conquests were aided by a little sleight of hand: the tricky lothario secretly placed cantharide beetles into his lady friends' clothes to boost their carnal appetites — he hoped they would literally be bitten by the "love bug." The story is undoubtedly apocryphal, but Casanova probably did experiment with Spanish fly.

We know that the Marquis de Sade did — he served cantharide beetles as dessert to party guests. The outcome, however, was not exactly the one he desired. An account of the adventure reads: "All at once, the guests, both men and women, were seized with a burning sensation of lustful ardor; the cavaliers attacked the ladies without any concealment. The essence of the cantharides circulating in their veins left them neither modesty nor reserve in the imperious pleasures; excess was carried to the most fatal extremity; pleasure became murderous. . . ."

Cantharidin's effects, far from being of an aphrodisiac nature, can be serious medical problems. In 1861, a French medical journal reported the most unusual case of several members of the Foreign Legion hospitalized in North Africa with prolonged and painful erections. The attending physician recognized the symptoms as those associated with cantharidin, but the soldiers denied experimenting with the substance. Indeed, their lack of access to female company seemed to corroborate the legionnaires' story.

Further questioning revealed that the affected soldiers had all dined on locally prepared frog legs. This gave the perceptive physician an idea: he went to the site where the frogs had been caught, found it to be crawling with cantharide beetles, captured a couple of frogs, and sacrificed them in the name of science. Their guts were filled with beetles. The doctor's curiosity was satisfied — surely the legionnaires had inadvertently ingested some cantharidin by eating the legs of those beetle-gulping frogs.

The theory, however, could not be tested because testing for trace amounts of cantharidin exceeded the chemical capabilities of the era. But, 130 years later, Thomas Eisner of Cornell University clearly demonstrated that cantharidin turns up in the legs of frogs that have been fed blister beetles; in fact, according to Eisner, people consuming a couple of legs from these frogs could be risking their lives. Just what we need — another thing to worry about. Now if we are going to dine on frog legs, we have to worry about what the frogs have been dining on.

Arthur Ford's tragic adventure began with an inquiry about the use of cantharidin as a treatment for warts. Pharmacists of his time stocked the preparation for this very purpose, as they do today. The skin irritation that cantharidin produces when applied topically does help eliminate warts, but if this stuff is taken internally it could eliminate the patient. This almost happened to a young Winnipeg girl who, for reasons known only to herself, swallowed a good dose of wart remover and ended up in hospital with a burnt esophagus and heart problems.

Knowing the dangers of cantharidin, I was intrigued by a mail-order advertisement for Spanish-fly candies. Curiosity got the better of me, and I sent off my 10 dollars; in return, I got five perfectly ordinary sweets, individually wrapped and neatly labeled "Spanish Fly Brand Candy." It seems there's a candy sucker born every minute.

ALICE IN MUSHROOMLAND

Charles Dodgson arrived at Oxford University in the mid-1800s as a young mathematics tutor, but by the time he died, some 50 years later, he had become the world-famous Lewis Carroll, creator of *Alice in Wonderland* and *Through the Looking Glass*. He wrote these delightful stories for a real live Alice, the daughter of Reverend Liddell, the dean of Christ Church College.

Dodgson made the fictional Alice the heroine of his tales and amused the real Alice by putting the character into incredible, whimsical scenarios. The writer did, however, draw on some real-life situations to create his Wonderland. A royal visit to Oxford, for example, inspired the Queen of Hearts; a traveling act called "The Talking Fish" inspired Carroll's own talking fish.

A book that Carroll is known to have read prior to writing these stories that captivated young and old alike is M.C. Cook's *A Plain and Easy Account of British Fungi*. This tome may well have inspired the caterpillar's remark to Alice as she contemplates eating the mushroom: "One side will make you taller and the other side will make you grow shorter." Did Carroll recall reading Cook on the hallucinogenic properties of the *amanita muscaria*, or *fly agaric* mushroom, when he penned these lines?

Indeed, some of Alice's adventures are so reflective of the effects of *amanita muscaria* that one wonders whether Carroll had more than a literary acquaintance with the mushroom. The reddish fungus with white polka dots is readily found in England; in fact, it is the mushroom most commonly depicted by illustrators of fairy tales. Who knows how many of these delightful stories were stimulated by mushroom power?

While ascribing Alice's strange adventures to mushroom

eating may prove controversial in literary circles, there is no question about the historical importance of mushroom indulgence. We can go back some two thousand years to a time when many generations of Indian medical tradition were formalized in the writings of the *Ayurveda*. These, in turn, were based on a much earlier collection of hymns known as the *Rig Veda*, dedicated in part to soma, a holy intoxicating beverage.

The opinion among ethnobotanists (scientists who study the relationship between plants and people) is that "soma" refers to preparations made from the amanita muscaria mushroom. The active ingredients in this mushroom, muscimole and ibotenic acid, are known to produce the kind of visions and euphoria described in the ancient writings. These works also refer to the ceremonial drinking by others of the urine of people who had consumed soma; the intoxicating constituents are known to be excreted unchanged by the body.

In the eighteenth century, a Swedish colonel held prisoner by a Siberian tribe described such a practice as engaged in by his captors: "Those who are rich among them, lay up large provisions of these mushrooms, for the winter. When they make a feast, they pour water upon some of the mushrooms, and boil them. They then drink the liquor, which intoxicates them. The poorer sort post themselves, on these occasions, round the huts of the rich, and watch the opportunity of the guests coming down to make water; and then hold a wooden bowl to receive the urine, which they drink of greedily, and by this way they also get drunk."

Similar stories are told about the use of amanita muscaria in Lapland, where the rich consume the mushroom at Christmas parties along with alcohol. When nature calls, they urinate in the snow, which is gathered and eaten by the poor looking for relief from their drab lives. Apparently, reindeer also love *amanita muscaria* mushrooms, and Laplander farmers scatter

pieces of them whenever they want to round up their herds. It is interesting to note that the culture of Lapland is replete with stories about flying reindeer, stories that probably gave rise to our Santa Claus fable. Perhaps those reindeer herders sampled some of their reindeer bait and actually saw their herds flying through the air.

The name "fly agaric" stems from the belief that the juice of crushed mushrooms will attract flies and cause them to fall into a stupor, rendering them ideal swatter targets. There is some question about whether *fly agaric* is truly effective in this way, and it may even be that the fly story was fabricated by people trying to conceal the real purpose behind their mushroom gathering.

There is a serious downside to eating amanita muscaria. Aside from hallucinogenic compounds, *amanita muscaria* contains the poisonous substance muscarine, which in high enough doses can kill. Dorothy Sayers used this as a plot device some 60 years ago in her celebrated mystery novel *The Documents in the Case*. A mushroom collector is found dead, apparently having mistaken *amanita muscaria* for the innocuous *amanita rubescens*. The victim's son refuses to believe that his father could make such an elementary mistake, and his tenacious investigation reveals that his father had in fact been poisoned with synthetic muscarine.

The amanita mushroom is a different species from the "magic mushrooms" that provoke such excitement among collectors every fall, causing them to trample many a farmer's field in search of ecstasy. The object of their search is the psilocybe mushroom, which contains the compounds psilocin and psilocybin, both capable of inducing hallucinations. This species was known to the Mayas and Aztecs and was widely used in religious ceremonies. When Montezuma was crowned in 1502, special priests presided over the use of teonanacatl, or "food of the gods," as the psilocybe mushrooms were known. The use of this hallucinogen was driven underground by Christian missionaries, but secret mushroom cults still exist in Mexico, where hallucinogenic fungi thrive on wet hillsides sprinkled with cow dung.

The lore of psychoactive mushrooms, as we have seen, is fascinating, and it seems that there is often more to these fungi than meets the eye. Sometimes, however, there is less. When Walt Disney's classic *Fantasia* was rereleased in 1991 in San Francisco, antidrug forces demonstrated in front of the theater claiming the film, with its dancing mushrooms, had a subliminal prodrug message. Silly stuff. I'm surprised these people aren't up in arms about *Alice in Wonderland*, but they probably aren't as up to date on Lewis Carroll's mycological interests as they could be.

CHEMICAL WITCHCRAFT IN SALEM

Could there be a more interesting place to visit on Halloween than Salem, Massachusetts? The town's stores are filled with witch memorabilia, witches are on hand to tell fortunes, and a sound-and-light show is mounted at the Witch Museum that grimly portrays one of the most disturbing events in American history.

The Salem witch trials of 1692 are among the best-documented witch-hunts. These tragic proceedings were initiated innocently enough when a few young girls began secretly to dabble in fortune-telling as a means of escaping the strictures of their Puritan existences. Their curiosity had been stirred by Tituba, a West Indian slave who entertained them with tales of black magic.

Everything was fun and games until one of the girls, who had devised a crude crystal ball of egg white, became convinced that she had glimpsed the image of a coffin within it. Soon, the other girls also began to have frightening visions, which led to panic fits, screaming, and bizarre behavior. The local physician could find no earthly explanation for the girls' apparent torment and concluded that they must have been bewitched.

The young ladies readily accepted this explanation since they were certainly not keen to reveal that they had been engaged in illicit fortune-telling. The hysteria spread, and soon people all over Salem began to show symptoms of being bewitched; the search for the witches responsible for all this suffering was on.

The afflicted girls, reveling in the spotlight, did not hesitate to point fingers. The scapegoats were stripped and examined for telltale "witches' marks," such as warts, which were supposedly used to suckle the devil. Even if no marks were found, the degree of hysteria manifested by the accusers during the questioning of the suspects could determine guilt. Before the madness was over, more than 200 people had been imprisoned for practicing witchcraft, 19 others had been hanged, and one had been crushed to death.

The Salem tragedy is usually described as a classic case of

mass hysteria. Some scientists, however, have offered an alternate explanation. It involves a fascinating ailment known as Saint Anthony's fire, an ailment from which the saint never actually suffered. A young, devout, third-century Christian, Anthony became disturbed by the ways of the world and decided to lead a simple life of seclusion in the Sinai desert. There, his loneliness grew, causing him to have hallucinations of wild animals and enticing girls, but in spite of these recurrent delusions he maintained his life of isolation and eventually founded the first Christian mission in Egypt. Anthony lived to the ripe old age of 105.

The moral strength he exhibited under duress appealed to those Christians who suffered from mental derangement. They often prayed to the saint, asking for help in coping with their problems, and evidently their prayers were sometimes answered. The ailment common to many of these people was characterized by disturbing hallucinations and a burning sensation all over the body. It came to be known as Saint Anthony's Fire.

Around the end of the sixteenth century, this disease was linked to the consumption of rye that had been contaminated with the ergot fungus (*claviceps purpurea*). Today, we understand that this fungus produces a variety of compounds (the ergot alkaloids) that can lead to convulsions, burning sensations, and the constriction of blood vessels. The latter symptom can lead to gangrene and the loss of fingers, toes, arms, or legs.

The active compounds in ergot have a chemical similarity to lysergic acid diethylamide, better known as LSD. In 1938, this powerful hallucinogen was produced from ergot by Albert Hofmann, a Swiss chemist. The ergot alkaloids themselves have been used in migraine medication and were once commonly administered to stop bleeding after childbirth.

How did praying to Saint Anthony cure ergot poisoning?

When those stricken with the condition made pilgrimages to the saint's shrines, they abandoned their usual diets, which included the contaminated rye. The monks in these shrines made bread from pure white flour, which soon came to be thought of as possessing curative properties. Today, there is no need to worry about eating rye bread because even if the grain has been tainted by the ergot fungus, modern milling techniques will eliminate it.

Now back to Salem. Rye flour was a dietary staple, and records show that the weather in 1692 was conducive to the growth of the fungus. The young girls, having low body weights, may have been among the most radically affected by the tainted flour; their fits of possession may have been induced by the mind-altering effects of various ergot compounds.

Curiously, one of the tests used to determine if the girls really were bewitched involved rye. Tituba was asked to bake a "witch's cake" with rye meal and the urine of the afflicted girls. This was then fed to a dog on the assumption that if the girls were truly hexed the dog would start showing the same kind of symptoms they had.

Sadly for modern science, the village minister did not accept this test as valid and the results were never recorded — the dog's behavior could have offered clues about the validity of the ergot theory. While the dog's strange antics may, in the seventeenth century, have been interpreted as evidence of witchcraft, modern chemical wisdom would have allowed for the alternate interpretation of an effect due to ergot alkaloids in the urine, but I guess we will never really know whether the inhabitants of Salem were victims of mass hysteria or of chemical witchcraft.

DEATH BY SOUFFLÉ

The caller on my radio show asked, "Why don't most doctors know about natural remedies?" Her words had a bitter and accusatory ring to them. I was sure I was in for another telephone diatribe about how some herbal therapy or dietary supplement had solved a medical problem that had stymied physicians. I was getting ready to respond with my usual spiel about the fallibility of anecdotal evidence and the need to examine such issues with scientific rigor when things took a more interesting turn: the caller informed me that she had finally found a doctor who had given her a list of foods to avoid and had thereby miraculously cured her. It was all so simple. Her problems had all been due to food allergies, she said, and she proceeded to list the foods she had to avoid: aged cheeses, chicken liver, pickled herring, chocolate, sausages, baloney, fava beans, Chianti wine.

At this point I realized that there was more to this story. I had just been given the classic list of foods that patients must avoid if they are prescribed a certain class of antidepressants known as monoamine oxidase inhibitors, or MAOIs. Apparently, an astute physician had diagnosed this woman's numerous and seemingly unrelated health complaints collectively as a sign of depression, had prescribed appropriate medication, and had given her good dietary advice. Perhaps a breakdown in communications had led to the patient's understanding that her problems were due to food allergies. But why are these strange dietary restrictions necessary for people taking MAOIs? Allow me to use a literary vehicle to tell you the whole story.

Rumpole of the Bailey is one of the most captivating characters in all of English literature. The creation of author John Mortimer, the cantankerous but likeable barrister simultaneously matches wits with London's criminal element and with "she who must be obeyed" — his wife, Hilda.

We cannot learn very much from Rumpole as to what constitutes the right chemistry for conjugal bliss, but by reading "Rumpole and the Expert Witness" we can certainly discover something about the consequences of marital discord. The story centers on a physician's clever plot to precipitate the untimely demise of his wife. The motive is an age-old one — another woman has entered the picture — but the method by which the rejected spouse is dispatched is novel: the murder weapon is a cheese soufflé. No, we are not talking about murder by cholesterol; the foul deed is accomplished with the nefarious use of a monoamine oxidase inhibitor.

A little background is needed at this point. In 1951, a new drug named iproniazid was introduced for the treatment of tuberculosis. It was one of the first effective pharmaceutical treatments for this dreaded bacterial scourge. When the drug proved to have adverse effects on the liver, it was replaced by isoniazid, a similar but safer product, which is still in use today.

Doctors, however, had noted a curious effect during the period that iproniazid was in use. Those taking the drug grew quite cheery. They were not exactly dancing in the halls, but there was no question that iproniazid had some sort of antidepressant effect. At first, little was made of this observation because depression theory of the time did not allow for chemical effects on the psyche, but the curiosity of several researchers was piqued, and by 1956 a dramatic experiment had been carried out. By injecting them with iproniazid, the researchers had induced euphoria in mice.

Stimulated by this result, Dr. Nathan Kline, a physician who had already made a name for himself by treating schizophrenia with a plant extract called reserpine, tried iproniazid on some of his depressed patients. The results were remarkable, and soon the drug was being widely prescribed as a mood-elevating substance. Yet it did not take long for doctors to become depressed about some features of the new antidepressant. While the medication worked well in most cases, there were unsettling reports of patients developing very high blood pressure and in some instances even suffering strokes. It was by using the medication to trigger this side effect that Mortimer's dastardly physician carried out his crime.

Extreme hypertension can occur when MAO inhibitors are combined with certain foods, drinks, or other drugs. Beer, wine, chocolate, chicken liver, pickled herring, and aged cheeses are most often associated with the hypertensive crisis. The inhibitors work their antidepressant magic by inhibiting an enzyme known as monoamine oxidase. This enzyme normally regulates concentrations of norepinephrine, dopamine, and serotonin, the brain chemicals that control our moods. Inhibiting the enzyme leads to higher levels of these substances and thereby alleviates depression.

The problem, however, is that the earlier-mentioned foods and drinks contain naturally occurring, blood-pressure-raising substances, such as tyramine, which under normal conditions are broken down in the body by monoamine oxidase. But if the enzyme is deactivated, levels of these chemicals rise and blood pressure can shoot up dangerously; that is why patients taking monoamine oxidase inhibitors today are given strict instructions about which foods they must avoid.

Now back to Rumpole. The physician had arranged for his wife to be prescribed a monoamine oxidase inhibitor for her depression. Then, as a treat, he made her a cheese soufflé and served it up with some wine. The tyramine in the cheese and wine conspired with the antidepressant to achieve the desired end: she succumbed to a stroke.

Far-fetched, you say? Not really. The medical literature records a number of cases of sudden death due to adverse reactions to monoamine oxidase inhibitors. In each instance, some substance that is normally metabolized by monoamine oxidase was introduced into the body. Since the enzyme was inhibited, an overdose reaction ensued. The painkiller Demerol, as well as cold remedies containing certain decongestants, have been implicated in such deaths. A plant-derived chemical, ephedrine, found in many "natural" weight-loss products as well as in the herbal supplement Ma Huang, can have catastrophic effects on someone taking an MAOI inhibitor.

In the Mortimer story, Rumpole does, of course, solve the crime and put the criminal away, but perhaps the real value of this tale is that it alerts people who do take monoamine oxidase inhibitors to the importance of avoiding certain foods and drugs. This may be Rumpole's best case for the defense.

MAD MONKS, KGB AGENTS, AND SLEEPING DOGS

Grigory Rasputin, the "mad monk," was a very powerful man in the court of Tsar Nicholas of Russia. This illiterate Siberian peasant with stringy hair, tangled beard, and a strong body odor established a position for himself in the Romanov court when he apparently saved the life of Nicholas's son Alexis, who was wasting away after sustaining a minor blow to the thigh.

Alexis was a hemophiliac, thanks to a gene he had inherited from his mother, Alexandra, the granddaughter of Queen Victoria. Rasputin told the tsarina that the way to save the boy was to keep him out of the clutches of the doctors and to pray. The advice turned out to be sound, because as soon as the doctors ceased their prodding and probing, Alexis's internal bleeding stopped. The tsarina and the tsar became indebted to Rasputin.

Rasputin's growing influence and bizarre behavior aroused a great deal of jealousy and concern among those at court. They frowned upon the monk's belief that to be absolved of sin, one must first commit a sin; the greater the sin, the greater the forgiveness when the sin was renounced. Indeed, if a young female penitent had not sinned enough, Rasputin was more than willing to help her.

The final straw for Rasputin's enemies was when the tsar began to take his advice on political matters. Under the leadership of Prince Yussopov, these courtiers hatched a scheme to eliminate the demonic clergyman who had already developed a supernatural aura after miraculously recovering from a stabbing. The conspirators would take no chances: they would poison him with cyanide. Yussopov enticed Rasputin to a party where he was served chocolate cake laced with potassium cyanide. The cleric ate and ate, but, to the horror of the onlookers, nothing happened. Was this fiend really possessed of supernatural powers? The conspirators panicked and one of them shot Rasputin in the chest, point blank. When Yussopov bent over to see if Rasputin had finally been dispatched, the "corpse" got up and began to chase him. Two more shots rang out and the monk finally slumped to the ground. He was then dragged outside and thrown into the Neva River, where, according to the autopsy, he finally drowned.

Why had the cyanide not done its job? After all, it is a notorious killer. Cyanide deactivates one of the most important enzymes in the body, cytochrome oxidase. This enzyme catalyzes the prime energy-producing reaction in cells — the one between glucose and oxygen — and when it is rendered inactive the body is left with no energy to run vital organs like the heart and lungs. Death rapidly ensues.

One possible explanation for the botched poisoning is that the schemers used old potassium cyanide that had become inactive, reacting over time with carbon dioxide from the air. Under these conditions, potassium cyanide slowly converts to potassium carbonate and releases hydrogen cyanide gas into the air. This theory is not as improbable as it may sound. Just a couple of years before the Rasputin incident, a Russian circus elephant had gone berserk and had had to be destroyed. The animal was very fond of cream cakes, and those charged with killing it decided to fill a hundred of these pastries with potassium cyanide. Although the elephant ate all of the cakes, it was unaffected — the unfortunate pachyderm eventually had to face a firing squad.

Usually, cyanide is a very "dependable" poison. That is why, during the Cold War, Soviet KGB agents relied on it to kill their political enemies. This came to light in 1957 when an exiled Ukrainian political leader and publisher of a Munich-based anti-Soviet newspaper was done in with cyanide. Some clever chemistry was involved. The KGB agent who was assigned the task of executing the publisher was equipped with a device that would generate hydrogen cyanide gas by mixing potassium cyanide with sulfuric acid. The gas could be directed at the intended victim's face, causing a rapid death that would be ascribed to a heart attack. (Incidentally, this is the same chemical reaction that was employed in the Nazi gas chambers, and it is still used in some American states where executions are carried out.)

Why was the KGB agent himself not affected by the gas as he carried out the assassination? He could have worn a gas mask, but this would not have been conducive to sneaking up on someone in public. There had to be another way. Soviet chemists had worked out an ingenious antidote system based on the body's mechanism for ridding itself of small doses of cyanide.

They knew that an enzyme called rhodanase converts cyanide to thiocyanate, which is excreted in the urine. This reaction, however, requires the presence of the thiosulfate ion, which is normally present in the body in very limited amounts.

The morning of the assassination, the agent consumed some sodium thiosulfate (also known as "photographers' hypo") for breakfast in order to prepare his body to handle the cyanide. Just before the fateful encounter, he crushed an ampoule of amyl nitrite in his mouth and inhaled deeply. This led to the synthesis of an altered form of hemoglobin called methemoglobin in his blood. Methemoglobin has a very high affinity for cyanide and binds the poison until it can be eliminated through conversion to thiocyanate as described earlier.

Although the chemistry is sound, this defense against the cyanide is not very reliable: the dose of the antidotes would have to be just right, since excess amyl nitrite is itself toxic. But it is interesting to note that the modern treatment for cyanide poisoning involves amyl nitrite inhalation followed by the intravenous administration of sodium nitrite (which also generates methemoglobin) and sodium thiosulfate.

This is exactly the treatment that was used several years ago on a Mexican medical student who, unable to wake up his sleeping dog, attempted mouth-to-snout resuscitation. It was all in vain — not only did the dog die, but also the student passed out. The attending physician at the hospital the student was taken to noted the odor of bitter almonds on the patient's breath and suspected cyanide poisoning. The dog had not been asleep; the animal had accidentally swallowed cyanide and had eliminated some of the poison through its lungs. Obviously, when it comes to cyanide-eating canines, it is best to let sleeping dogs lie.

CHEMISTRY FOR ZOMBIES

A few years ago, Michael Jackson thrilled the world with his video portrayal of a singing zombie. Can such things be? Can the dead rise from their graves and prowl the countryside, terrifying the living? What better place is there to check out such stories than Haiti, the land of voodoo and zombies.

Movie audiences have long been terrorized by tales of the living dead, but it took a Harvard scientist to unearth the science behind the myth. Wade Davis explored the backwoods of Haiti in an attempt to discover the origin of the zombie stories and, apparently, came across a real live specimen.

Clairvius Narcisse, a poor Haitian peasant, had died in 1962 and was appropriately buried, but 18 years later he startled his sister by showing up in the local marketplace very much alive. He told her that their brother, with whom he was involved in a land dispute, had employed a voodoo priest to turn him into a zombie. After his burial, he was resurrected and forced to work with other "zombies" until he managed to escape two years later. He then wandered the country, fearful that his brother would recognize him. It was only after he heard about the brother's death that he decided it was safe to come back to life.

Narcisse told of being rubbed with a "zombie powder," inducing a deathlike state, and of being "resurrected" after the effects wore off. Then, he maintained, he was kept in a perpetually drugged condition to prevent him from escaping. Wade Davis was prompted to investigate this fascinating story by a psychiatrist who was interested in potential medical applications of the "zombie powder" — if such a substance existed. The psychiatrist, Dr. Nathan Kline, had already made a name for himself by using reserpine, a drug isolated from the Indian snakeroot plant, in the treatment of psychiatric patients. He wondered whether the active ingredient in zombie powder might also have medicinal properties. Equipped with a camera and some money, Davis was able to track down a number of voodoo sorcerers who claimed to be able to produce — for a fee — the so-called zombie powder. Several times, Davis witnessed these practitioners mixing ingredients ranging from the crushed skulls of freshly exhumed babies to extracts of various toads, but the only ingredient that appeared common to all the preparations was a particular type of fish known as the puffer or blowfish.

This is where things started to get interesting for Davis. He knew that the liver and sex organs of the fish contained a poison, known as tetrodotoxin, that could paralyze the nervous system. He was also aware that a number of Japanese gourmets who had dined on improperly prepared puffer fish, or "fugu," had died as a result — although fugu chefs receive extensive training in the extraction of dangerous organs, they occasionally err and kill their customers. But there was one stunning account in the medical literature of a fugu-poisoning victim who suddenly sat up as he was being wheeled into the morgue: a prototype zombie.

In a state of great excitement, Davis returned to the United States and had samples of the zombie powder analyzed. Sure enough, a couple showed the presence of tetrodotoxin. Further encounters with the Haitian sorcerers yielded another concoction, made from "zombie cucumber," which was alleged to keep resurrected zombies in a state of stupefaction. The "zombie cucumber" turned out to be none other than *datura stramonium*, replete with psychoactive atropine and scopolamine.

Both atropine and scopolamine can cause disorientation, confusion, amnesia, stupor, and bizarre behavior — decidedly zombie-like symptoms. Scopolamine has actually received attention as a "truth serum" because of its ability to disorient and to induce a sedated condition. The theory is that in a sco-polamine-induced state, a subject doesn't have enough mental energy to concoct a lie. Blurred vision and difficulty in main-taining equilibrium are also scopolamine reactions. Now, doesn't that conjure up an image of crazed zombies stumbling around the countryside?

So, Davis seems to have a case, and his arguments would appear to be buttressed by a report published in the British medical journal the *Lancet* of a Singapore man who ate a blow-fish and fell into a coma for 36 hours. During this period, he had no brain-stem reflexes, which usually means extensive brain damage, yet this man — who, to the casual onlooker, would have appeared quite dead — recovered completely within a week.

Davis recounts his whole fascinating adventure in a book entitled *The Serpent and the Rainbow* (which was made into a rather poor movie). Along the way, the zombie theory has somehow become fact — but hold on: the scientists who ana-lyzed the zombie powder maintain that while they did find tetrodotoxin, the amount was so small that it could not have induced the zombie-like state. Davis, however, insists that the presence of any tetrodotoxin means that other samples may contain far more than the ones he had been able to secure.

There is no question that tetrodotoxin does interfere with nerve function. Even the mechanism of the action is under-stood; the drug blocks the uptake of sodium by cells, a process that is critical to the transmission of signals from one cell to another. It is intriguing that according to voodoo legend, zom-bies are not allowed to eat salt, lest they be "dezombified." Salt is sodium chloride. Might sodium neutralize the effect of the tetrodotoxin? In any case, there you have it — the whole story of puffer fish, zombie cucumbers, and the living dead. Interest-ing . . . but let's take it all with the proverbial grain of salt.

HEALTH AND DISEASE

SOLA DOSIS FACIT VENENUM

Philippus Aureolus Theophrastus Bombastus von Hohenheim. With a name like that, it is little wonder that this extraordinary, albeit outrageous, sixteenth-century Swiss healer preferred to call himself Paracelsus. This was not exactly a modest alias, deriving from the name of Celsus, one of the most famous of the ancient Roman physicians, and from the Latin word "para," meaning "beyond." Paracelsus evidently considered himself superior to the doctor on whose writings Renaissance medicine was based — rather arrogant for someone who, according to historical records, called himself "doctor" without ever having completed the required formal training.

In fact, Paracelsus despised the universities and their graduates. He challenged their teachings, calling them antiquated and abhorring their uncritical reliance on ancient authorities. He claimed that doctors, instead of curing their patients, either killed or lamed them with their purgatives, their bloodletting, and their complex plant medications. Vitriolic in his attacks against the establishment, Paracelsus lashed out at physicians, saying that "there are some who have learned so much that their learning has driven out all their common sense, and there are others who care a great deal more for their own profit than for the health of their patients."

Paracelsus offered an alternative to what he perceived as crude attempts at healing: "The universities do not teach all things, so a doctor must seek out old wives, Gypsies, sorcerers, wandering tribes; a doctor must be a traveler . . . knowledge is experience." To emphasize his views, as the legend goes, he stood one day before a crowd of cheering students and burned the books of Avicenna and Galen, perhaps the best-recognized medical authorities. This is probably an embellished account, because in those pre-printing-press days, handwritten volumes such as these would have cost a fortune.

There is no doubt, however, that Paracelsus's outlandish antics won him a large following. His fame spread; his lectures were very well attended. He spoke out against useless potions and infusions and stressed the healing powers of nature. The answer to many medical problems, he maintained, lay in harnessing the developing science of chemistry. Substances isolated from nature, be they of a plant or mineral origin, could cure disease, but these cures could only be discovered through experimentation — out with reliance on the words of ancient doctors, and in with laboratory and clinical experiments.

Disease is a localized abnormality, not an imbalance of humors, Paracelsus maintained. It is a chemical problem to be chemically treated. For this insight alone, Paracelsus may be identified as the father of modern pharmacology. He proclaimed that alchemists should extend their goals beyond converting metal to gold and broadened the definition of alchemy to include any process in which a naturally occurring material is transformed into a new substance: "For the baker is an alchemist when he bakes bread, the vine-grower when he makes wine, the weaver when he makes cloth." But there is no doubt that Paracelsus believed the most important use to which alchemy could be put was to prepare medicines.

Adhering to his own doctrine, Paracelsus traveled widely, picking up valuable knowledge and experience. He became convinced that specific diseases needed treatment with specific drugs, not with the mishmash of plant ingredients that the apothecaries prescribed for virtually all conditions. "The apothecaries are my enemies," he stated, "since I refuse to empty their jars. My prescriptions are simple and have no need of forty or fifty ingredients. I aim not to make apothecaries rich, but to cure patients." And, sometimes, that's just what he did. "I please nobody except the people I cure," he taunted his critics.

Paracelsus introduced laudanum, an extract of opium, for the treatment of pain and popularized mercury compounds as a therapy for syphilis. He was probably the first person to emphasize the importance of drug dosage, and he railed against the overuse of mercury. His often-quoted comment, "*sola dosis facit venenum,*" usually translated into English as, "all medicaments are poison and only the right dosage makes them stop being poison," is considered to be the cornerstone of the science of toxicology.

Just as important was Paracelsus's recognition of the potent link between the body and the mind. "A doctor's personality can act more powerfully on a patient than all the remedies he prescribes," he would often say. When his treatments failed, he would sometimes dispense a mysterious powder from his hollow sword handle and, with a flourish, administer it to the patient. This placebo treatment often had a near-miraculous effect, supporting Paracelsus's theory that "imagination takes precedence over all."

Paracelsus further realized that the patient must be treated as a whole: diet, exercise, surroundings, and even massage are as important as specific chemical remedies — advanced thinking for someone who was condemned as a charlatan by the medical authorities of his day.

Although Paracelsus was a remarkable visionary, he was still a product of his times. He pioneered the use of ether as an anesthetic three hundred years before its widespread use, yet he maintained that the devil created all insects from menstrual blood. He discovered that anemia sufferers responded to iron salts, yet he taught his students to treat wounds by rubbing a special ointment on the weapons that had inflicted them. While he believed that diseases were abnormalities that could be treated with simple chemical remedies, he was convinced that the planets somehow corresponded to parts of the body.

Obviously, some of Paracelsus's ideas were just plain balderdash, but when these are filtered out, we are left with a remarkable legacy. More than any other single person, this medieval alchemist was responsible for teaching the importance of relying on observations rather than on ancient authorities, for uniting chemistry with medicine, and for introducing the idea of specific chemical remedies for diseases. He emphasized the crucial relationship between body and mind with his thoughtful, although overly optimistic, observation that "He who is happy always gets well."

Above all, we remember this complex man for introducing us to the idea that only dosage determines the difference between poison and cure. We should reflect on this dictum whenever questions about the toxicity of pesticides, water pollutants, or food contaminants arise, as well as when we evaluate the potential benefits of eating broccoli, taking dietary supplements, or drinking tea. Of course, Paracelsus also had his faults: he often prescribed "*zebethum occidentale*," or dried human excrement, for the treatment of sore eyes. I doubt it did any good. Did it do harm? Well, *sola dosis facit venenum*!

ANXIETY ABOUT ANXIETY

There's a fascinating legend that South Pacific islanders pass down from generation to generation about a supreme being called Tagaloalagi, who created the earth and everything on it. This included the first man, Pava. Tagaloalagi was pleased with his work and celebrated by sitting down with Pava to share a beverage from the roots of a sacred plant. The space between the drinkers was hallowed, Tagaloalagi decreed, and it could not be intruded on until the ceremony was over. Pava's young son violated this rule, and both he and Pava were severely reprimanded by Tagaloalagi. When the boy transgressed a second time, the Divine One lost his temper and tore the boy limb from limb. Pava was devastated. He had lost his only son. How would the world be populated now? When Tagaloalagi saw that Pava recognized the boy's sin, he declared that the punishment, while just, would not be permanent. He spilled a few drops of the sacred beverage on the boy's body, and he was immediately restored to life. For South Seas natives, that drink, known as kava, has come to symbolize the link between man and his creator. To this day, throughout the Pacific, kava ceremonies begin with the spilling of a few kava drops on a mat.

Kava is consumed to celebrate marriages, to welcome visitors, to aid decision making, and to mourn the dead. Some proponents even claim that consuming the beverage facilitates communication with the departed. Kava undoubtedly does affect the brain, and those who overdose on it may indeed hear voices — it is unlikely, though, that these voices come from beyond the grave. Kava is also commonly used as a social lubricant — the islanders' version of the five-o'clock martini.

Part of kava's attraction derives from the curious fashion in which the beverage was traditionally prepared. The raw materials needed were the *Piper methysticum* plant and a few virgins.

These ladies chewed the root of the plant and spit the resulting mash into a communal pot. This unappealing concoction was then diluted with water and consumed to produce a feeling of contentment and relaxation. To this day, on islands like Fiji, the lobbies of banks and business establishments feature public kava bowls. Virgins, however, are no longer involved in the preparation of the beverage.

Why are we talking about the quasireligious practices of Polynesians these days? Because kava has crossed the Pacific and is rapidly becoming one of the hottest commodities in North America. It is being promoted as the herbal Valium, the natural Xanax. The solution to chronic anxiety in North America. Delight without danger. Relaxation without risk.

We are used to hearing all kinds of herbal hype these days, most of which stands on pretty shaky scientific ground, but kava really does have a measurable physiological effect. While it is no panacea, it may prove to be a useful antianxiety agent and sleep promoter. The active ingredients in the plant's roots are compounds known as kavalactones. Researchers have shown that these compounds enhance the activity of a neurotransmitter in the brain known as GABA (gamma aminobutanoic acid), which has been linked with feelings of mellowness. Kavalactones are soluble in water and consequently are consumed in the form of an infusion made by grinding the root and mixing it with water. The finer the grind, the more easily the active ingredients dissolve. Chewing the root produces a fine mash that releases the kavalactones very easily. This explains the tradition involving virgins: chewing is important; sexual history is not.

Kava products are now available in liquid, capsule, and tablet forms. Many kava preparations, to their manufacturers' credit, list the weight of a dose along with the percentage of kavalactones it contains. A standard label might say, for example, that each pill weighs 250 milligrams and contains 30 percent kavalactones; this means that a tablet has 75 milligrams of the active ingredients. But whether the tablet actually contains what the label says it does is another matter. Furthermore, different *piper methysticum* plants have different relative compositions of kavalactones, and, in any case, no one really knows which lactones are the most desirable.

Quality studies on kava dosage are scarce, but a few intriguing ones have been carried out in Germany, a country that has a long tradition of investigating herbal treatments scientifically. Several have shown that a person's anxiety levels can be reduced in a week if about 70 milligrams of kavalactones are taken three times a day. One trial compared the effect of kava to that of Serax (oxazepam), a commonly used antianxiety agent. Although the effect was about the same, kava produced none of the side effects — like drowsiness, dizziness, headache, or vertigo — sometimes associated with the prescription drug. Subjects stated that their minds felt completely clear on kava, and they did do very well on word-recognition tests. Neither were there any problems associated with stopping kava; people who stop taking the common antianxiety benzodiazepines, like Valium, sometimes experience withdrawal symptoms ranging from insomnia to psychosis. Still, not enough is known yet about kava to consider it as an alternative to antianxiety medications, which have a proven track record. It may, however, be appropriate for a physician to prescribe kava in cases of mild anxiety before resorting to standard drugs.

Kava also holds promise as a treatment for sleeplessness. This isn't surprising, since the problem is often the result of anxiety. To promote sleep, a dose of 150 to 200 milligrams of kavalactones taken a half hour to an hour before bedtime appears to be effective. At these doses, side effects are virtually nonexistent. But kava can be abused: ingesting excessive amounts can lead to a loss of muscle coordination, producing a state that resembles alcohol inebriation. In Utah, a motorist was actually convicted of impaired driving even though his blood-alcohol level was zero; he had admitted to drinking 16 cups of kava. Certainly, kava should not be combined with alcohol or any other antianxiety medication.

In rare cases, high doses have even triggered involuntary muscle movements, raising the question of whether kava blocks the activity of certain neurotransmitters, such as dopamine. Since this substance is in short supply in the bodies of those suffering from Parkinson's disease, kava is probably best avoided by Parkinson's patients. A yellow skin discoloration accompanied by scaling has also been noted as occurring at abusive doses. The longterm effects of daily use have not been studied, and it would therefore be wise to limit a trial of kava to about three months. For similar reasons, pregnant or nursing mothers should not experiment with kava.

The occasional use of kava appears to be completely innocuous. "Why not take a swig, or try a capsule?" say its advocates. A recent preliminary study has even shown that kava may be capable of reducing the stresses associated with everyday living, such as in-law visits, spousal arguments, or car trouble. Food producers, attempting to capitalize on the public's anxiety about anxiety, have started to add kava to snack foods. Now, I'm not averse to a little tranquility, so I bought a bag of kava-laced corn chips. Instead of relaxing me, it increased my stress level because there was no mention at all about how much kava the chips contained. What I needed was a standardized kava capsule to calm me down. I took one about an hour before writing this piece . . . I can't say I feel a significant effect, but I think it has made me more relaxed, more carefree, and perhaps less critical. I probably wouldn't have written as positive a commentary on kava without it.

Colorful Wastes

It sure is a colorful life. I recall vividly a scary episode a few years ago when my wife came running to me brandishing our infant daughter's diaper. It was filled with bright red stuff. After a few moments of panic, we took a closer look and realized that earlier in the day the young lady had been introduced to the delights of red licorice — the red dye was making a triumphant exit.

This is, of course, not a unique experience. The scientific literature records the case of a young boy who succeeded in terrifying his mother with a bright red-orange output after dining on some colorful Nerds cereal. More unusual is a condition called "Hydrox fecalis," aptly named after a brand of chocolate sandwich cookie. These sweet treats are colored with cocoa powder, which can turn the stool black approximately 18 to 24 hours after ingestion. For this to happen, however, you do have to eat a lot of cookies — like half a pound. This is enough to cause abdominal pain, which, coupled with the black color, can certainly throw a good scare into someone.

Black stools are a real cause for concern because they can be a sign of gastrointestinal-tract bleeding, but they can also be caused by iron supplements, black licorice, blueberries, or medications that contain bismuth compounds (such as Pepto-Bismol). Ignorance of the blueberry effect has resulted in many an unnecessary trip to the emergency room.

The brown color of normal feces is mostly due to remnants of bile secreted by the liver into the small intestine. It is also due to bilirubin, a major breakdown product of red blood cells. Bilirubin's precursor is a green compound known as biliverdin; this compound sometimes shows up in the feces, making them green. This happens when transit time through the digestive tract is rapid (often due to a viral infection) and there is a reduced opportunity for biliverdin to be converted to bilirubin. Babies will often have quick transit times and produce bright green stools.

Unusually light or clay-colored feces can be a sign that there is a blockage in the bile duct. This is very uncommon — when it does occur, it's more likely the patient has consumed a large amount of white Maalox or other antacid.

Some of the bilirubin is absorbed into the bloodstream and eventually excreted in the urine, giving it its yellowish tint. When the kidneys secrete considerable water, the urine is pale, but when the body needs to conserve liquids, the urine is more concentrated and therefore a darker yellow. After a person has performed some heavy, sweat-producing exercise this often happens; very dark yellow urine can be a sign of dehydration.

If too many bile pigments end up in the urine due to impaired liver function, the urine takes on a greenish appearance, although this can also be the result of consuming asparagus — some people possess the particular genetic trait that causes this to occur. Similarly, about 15 percent of the population will produce red urine after eating beets. This latter phenomenon is a particularly interesting one. Beetroots have both red (betacyanins) and yellow (betaxanthins) pigments, known collectively as betalains. The betacyanins, of course, greatly exceed the betaxanthins. Dried beet powder is available commercially and has been used to color candies, yogurts, ice creams, salad dressings, drinks, and gelatin desserts.

Most people do not have to deal with the aftereffects of beet consumption because hydrochloric acid in their stomachs and bacteria in their colons break down the pigments before they can be absorbed into the bloodstream. But not everyone has the same mix of colon bacteria — those who experience a colorful beet aftermath are, apparently, missing the bacteria that degrade the betalains.

Under certain conditions, people who have not previously had the experience may suddenly start noticing a red tinge to their feces or urine after eating beets — it all depends on what else they ate with the beets. Oxalic acid, which is found in a variety of foods, actually protects betalains from being broken down by bacteria. Oysters, spinach, and rhubarb, when eaten with beets (admittedly an odd combo), can have an effect that is disturbing for anyone not familiar with this bit of obscure chemistry.

The occurrence of red urine is understandably very frightening, because the discoloration may be a sign of the presence of blood and can indicate kidney or bladder problems. It must be investigated; in some cases, its cause will be an innocent one. For example, drummers can have red urine because they repeatedly tap their fingers and hands, breaking red blood cells which then release their hemoglobin into the urine; in some parts of Africa, the contention is that if a drummer doesn't pee red he is not playing well.

Purple urine may be a symptom of porphyria. In one variant of this disease, an inherited enzyme deficiency interrupts the metabolic pathway for the production of hemoglobin. Porphyrins, which are molecules that would normally be used by the body to make hemoglobin, are then excreted and appear in the urine. Porphyria can have various symptoms, including mental impairment. King George III, the British monarch during the American War of Independence, is believed to have suffered from porphyria, and this perhaps explains his bizarre handling of the colonial situation.

It is no wonder that doctors — or, in the Middle Ages, specialists known as "piss prophets" — have historically taken pains to examine the color of urine for signs of disease. I do wonder, however, how many people were subjected to nonsensical "medical" treatments as a consequence of having eaten beets or asparagus. One final colorful bit of lore: according to an old wives' tale, if you want to know whether a baby will be a boy or a girl, take a sample of urine produced during the sixth month of pregnancy and mix it with an equal volume of liquid Drano; if the mixture turns green, it's a boy, and if it turns yellow, it's a girl. The test is correct about 50 percent of the time.

Bee Pollen and the Office of Alternative Medicine

Why bother going to see a doctor if you feel sick? Just walk into any bookstore these days and check out the health section: there you'll find a cure for everything. If you have asthma, just rub oil of oregano on your chest. Digestive problems? You need the Clay Cure. Magnets will relieve your arthritis pain, and aromatherapy is the solution for ailments ranging from cystitis to anxiety. Then there are the nutritional regimens. Depending on which book you leaf through, salvation lies in flaxseed, fish oils, garlic, oat bran, soy protein, red wine, freshly squeezed juices, apple cider vinegar, or barley green powder. And let's not forget those supplements — vitamins, pycnogenol, blue-green algae, tea-tree oil, bifido bacteria, natural enzymes, and shark cartilage will come to the rescue.

If none of this entices you, then try drinking some tea in which the revolting, slimy "kambuchia mushroom" was grown, or experiment with colorpuncture, a technique that focuses colored light on acupuncture points and "energizes powerful healing impulses." Uri Geller's Mind Power Kit will assist you to use crystal quartz for psychic healing, and books on feng shui will teach you how to harness positive energy from the environment through the correct placement of furniture and decorative items in your home or workplace. You can also discover the secrets of holistic bathing (whatever that is), chelation therapy, bee pollen, homeopathy, Ayurvedic medicine, natural hygiene, chiropractic, catalyst-altered water, colonic lavage, therapeutic touch, coffee enemas, and naturopathy. Confused? Uncertain about what to do? Just pick up a book on flower remedies and discover that scleranthus extract is the answer to vacillation, indecision, and uncertainty.

What is the common feature of all of these therapeutic approaches? They are all propped up with piles of anecdotal evidence. Breast cancer victims describe how their breast lumps disappeared after they went on an organic-juice diet, and indigestion sufferers tell us how they found relief after ridding their intestines of "parasites" with some miraculous herbal concoction. It all sounds great. The only thing missing is scientific evidence. There are no controlled studies to back up the claims, no follow-up investigations to see whether the reported cures were maintained. Of course, a lack of controlled studies does not mean that a particular treatment does not work; after all, many medical discoveries have started with anecdotal evidence. Someone may make an observation that initially appears outlandish — like the observation that eating limes prevents scurvy. In 1754, James Lind, a Scottish physician, was ridiculed for suggesting that sailors be given citrus fruits on long sea voyages to prevent this dreaded disease. Soon, however, he was vindicated: sailors who adhered to the usual diet of dried bread and salted meat got scurvy, whereas those who supplemented this meager fare with limes did not. Anecdotal evidence was transformed into scientific fact.

It is certainly possible that some of the remedies and treatment regimens that vie for bookstore shelf space will undergo the same transformation, but until that happens, they will remain in the category that we have come to refer to as alternative therapies. This does not necessarily mean that they are ineffective, only that they are untested or unproven. What is an established fact is that people are scrambling to obtain these therapies. Modern scientific medicine cannot provide cures for all ailments, and in many cases physicians are perceived as beleaguered, uncaring, and unaccepting of new ideas. Alternative practitioners are usually charismatic, they spend a great deal of time with their patients, and they insist that there's a good chance they can help. They offer hope, although often it turns out to be false hope.

What we really need is a thorough scientific examination of the alternative therapies that show promise based on personal testimonials. This process is starting to take shape. In the United States, the Office of Alternative Medicine was created in 1991 and given a budget of two million dollars; it has since become the National Center for Complementary and Alternative Medicine and will have a budget of fifty million for 1999. The center will award grants and organize clinical trials. Perhaps we can look forward to some interesting results, but so far, since 1991, not much has happened: not a single "alternative" treatment has been proven highly effective, and not a single one has been completely debunked.

It is surprising that the Office of Alternative Medicine never thoroughly investigated the potential of bee-pollen therapy, since that therapy played a crucial role in the office's creation. While the force behind the establishment of the Office of Alternative Medicine was Senator Tom Harkin of Iowa, the spark was provided by another Iowa politician, Berkley Bedell, who became thoroughly taken with alternative medicine when he apparently cured himself of Lyme disease and prostate cancer by ingesting colostrum, the first milk of a cow that has just given birth. This treatment was the brainstorm of Herb Saunders, a Canadian farmer who, for $2,500, would sell a sick person a pregnant cow, inject some of that person's blood into the cow's udder, and then provide him or her with the colostrum. He claims that the colostrum has "the power to wham out cancer." The authorities don't agree — Saunders has been twice arrested for swindling, mistreating animals, and practicing medicine without a license. Colostrum did seem to cure Bedell, however, who then contacted Harkin in his pursuit of ways to fund alternative medicine.

During their discussions, the subject of Harkin's allergies came up. Bedell, already fancying himself an alternative-care expert, suggested to Harkin that he try bee pollen. Harkin began to take pollen tablets, sometimes up to 60 a day, and claimed that after six days his allergies disappeared. Understandably very impressed, Harkin immediately began to lobby for the establishment of the Office of Alternative Medicine. Today, he says he still sometimes suffers from allergies, but when they manifest themselves he just takes more pollen and they disappear. Still, there have been no corroborating studies: anyone contemplating using bee pollen should remember that in rare cases it has triggered life-threatening allergic reactions.

Even though the cures that stimulated the creation of the Office of Alternative Medicine are suspect, the National Center for Complementary and Alternative Medicine can certainly go on to serve a useful purpose. Scientific investigation of claims for alternative medicine is sorely needed. It may turn out that colostrum actually has beneficial properties — some studies have shown that cows may, in fact, form useful antibodies to injected microorganisms. Even bee pollen may turn out to be useful. But what we need are facts, not hype. In the meantime, I still despair when I walk through the health section in a bookstore, because I wonder how many people are unsuccessfully trying to restore their health by "rolfing" or undertaking auto-urine therapy. Maybe I'm just being crabby. Maybe what

I need is some extract of *malus pumilia*, also known as crab apple, which — according to *The Bach Flower Therapy Book* — reduces despondency and increases broad-mindedness.

FIGHT CRIME: EAT CHALK

One of the most arresting demonstrations I perform in the lecture room, or so my students tell me, is eating a piece of chalk. I normally do this when we discuss the chemistry of calcium supplements, pointing out that the source of calcium carbonate is irrelevant. Naturally, I am not the first person to take an unusual calcium supplement: that honor belongs to Cleopatra.

Sometime during the first century B.C., the Egyptian queen bet her lover, Marc Antony, that she could invite him to the most expensive dinner ever served. Marc Antony had enjoyed some rather elaborate meals in his time, so he agreed to the wager. When the appointed hour arrived, he sat down at a table set with nothing but a goblet containing a clear liquid. As Antony's anticipation grew, Cleopatra carefully removed one of her huge pearl earrings, crushed it, and dropped the powder into the goblet. The liquid, which was actually vinegar, fizzed impressively as the bits of pearl dissolved. The queen picked up the goblet and triumphantly drank the potion. She had indeed consumed the most expensive dinner of all time: the pearl was worth as much as two million ounces of silver. By committing this act, Cleopatra may also have become the first woman ever to make use of dietary calcium supplements.

Pearls are essentially composed of calcium carbonate, the active ingredient in many calcium tablets used today. These supplements can help prevent the bone-brittling disease known as osteoporosis. But increased calcium intake may have even more wide-reaching effects. It may offer protection against

kidney and colon cancer; furthermore, calcium increases the rate at which the body produces nitric oxide, a chemical instrumental in relaxing the walls of blood vessels, thereby lowering blood pressure.

And, believe it or not, calcium consumption may even reduce the crime rate. At least one study has linked high blood levels of lead and manganese with murder, assault, and robbery. Researchers suggest that these minerals are absorbed into the brain far more readily if there is an inadequate calcium intake. Make those criminals drink milk!

Osteoporosis is a serious disease, striking about one-quarter of all women over the age of 50 and causing about 350,000 hip fractures a year in North America. Roughly 15 percent of these hip-fracture victims eventually die of circulatory problems, blood clots, or pneumonia — all well-established complications of such injuries. Broken wrists and loss of height due to fractures of the vertebrae can also be direct results of osteoporosis.

Low calcium intake is not the only predisposing factor for the disease: too much protein and salt in the diet, too little vitamin D, too little exercise, early menopause, longterm cortisone therapy, and smoking are just some of the other risk factors. But increased calcium intake is something that most people can readily and safely accommodate to increase bone strength.

Bones derive their strength from a matrix of flexible protein fibers combined with hard calcium phosphate crystals. These crystals, however, are not static; bone is living tissue that is constantly being "remodeled." This term simply means that there is a constant turnover of bone, with some minerals being deposited to form bone and some dissolving from bone into the blood. This latter process is known as "resorption."

Calcium, the most abundant mineral in the human body, serves a variety of functions in addition to playing a role in bone formation. It is essential for blood clotting, for the normal functioning of nerve tissue, and for the contractions of smooth muscle. Even the beating of the heart is regulated by calcium levels.

Since blood calcium is essential to life, the body will attempt to maintain adequate levels even at the expense of bone resorption. If the bones are well formed and contain enough calcium, there is no problem; but if there is insufficient bone mass, osteoporosis and all of its consequences can result.

How do we know what the ideal calcium intake is? An important clue may be obtained by measuring calcium output in the urine: when one's intake is greater than about one thousand milligrams, the calcium concentration in the urine increases, meaning the body has retained as much as it needs. It seems that a thousand milligrams of calcium a day for premenopausal women and men below the age of 65 is appropriate to achieve a calcium balance, but about fifteen hundred milligrams are needed by men over the age of 65 and postmenopausal women who are not taking estrogen supplements. Taking estrogen reduces the requirement to around one thousand milligrams.

The best sources of calcium in the diet are dairy products; in fact, it is difficult to meet one's daily calcium needs without them. A glass of milk has about three hundred milligrams, and a cup of yogurt has four hundred. By comparison, the best vegetable source is broccoli, with approximately one hundred milligrams per cup. Unfortunately, some people have been frightened away from milk products because of concerns about increased blood cholesterol and the unfounded allegation of some activists that "cow's milk is for calves, not humans."

Yes, full-fat dairy products can increase blood cholesterol levels, but these products are easily avoided. Today, a wide variety of low-fat and fat-free products containing the same amount of calcium as their higher-fat counterparts are available. Fluid milk also contains vitamin D, which is essential for proper calcium absorption; alternatively, a 15-minute exposure to sunlight daily can generate enough of this vitamin. Calcium-fortified orange juice containing 350 milligrams per cup has also appeared on the market, making it easier for consumers to achieve their dietary-calcium goals.

Still, many people find it difficult to consume a thousand milligrams of calcium a day, and they resort to supplements. But how do they decide which one to buy? Actually, the form calcium comes in appears to be of little significance, but to derive the maximum benefit from the mineral it is important to get enough exercise.

Calcium lactate, calcium gluconate, calcium citrate, and calcium carbonate are all suitable, and they are best taken with meals. Calcium citrate may be somewhat more readily absorbed, but it contains less calcium than calcium carbonate — 24 percent by weight compared with 40 percent. Remember that dietary recommendations are always in terms of calcium alone, which makes up only part of the weight of a supplement. Calcium carbonate is therefore the most efficient source, although it may have a slight constipating effect.

As far as the body is concerned, it makes no difference whether the calcium carbonate is manufactured in a laboratory or comes from pearls. Whether one chews on Tums, grazes on the White Cliffs of Dover, or dines on chalk is purely a question of personal preference.

I usually end my chalk-eating lecture by asking the students to estimate their daily calcium intake. The results are almost always frightening — many have intakes below four hundred milligrams. While they are very hesitant when I offer them a bite of chalk, I know they've bought my arguments when I see them happily drinking the skim milk I pass around. Maybe it doesn't have the greatest taste, but it will go a long way towards reducing their risk of osteoporosis and perhaps even heart disease. And it sure beats eating chalk.

FEELING NO PAIN

Freedom from pain is probably the single most important criterion for happiness. An intense toothache, for example, makes us forget everything else; neither election results nor the vagaries of the stock market hold any interest: we just have to get rid of that pain. So off we go to the dentist, who administers an anesthetic of some kind and proceeds to solve the problem by drilling, cutting, or pulling. Life is worth living again, and our interest in the world is reborn.

Can you imagine enduring this without benefit of a painkiller? Worse than that, can you imagine having a leg amputated or a gallstone removed without anesthesia? Yet this is exactly what patients had to endure prior to 1846, a pivotal year in the history of science. Alcohol was available, but no matter how drunk a patient was, he'd still feel his leg being sawed off.

Towards the middle of the nineteenth century, chemistry came to the rescue. Before it could do this, a process of discovery had to take place. Joseph Priestley, the brilliant, self-educated British chemist, created what he described as a novel "air" by carefully heating ammonium nitrate. "Carefully" is the operative word, because ammonium nitrate can explode. Priestley's "air," which turned out to be nitrous oxide, came to the attention of young Humphrey Davy, who, as a 17 year old, had begun experimenting with it, noting its intoxicating effects — it actually made people giggle. Davy went on to become a noted scientist. He recorded his observations in a book, published in 1800, even mentioning how the gas had relieved his own headaches. Soon the word was out, and laughing-gas parties became popular, particularly among students and intellectuals. But nitrous oxide was not the only intoxicating vapor that fueled these parties.

Ether had been made by the Prussian botanist Valerius Cordus in 1540 by reacting sulfuric acid with alcohol. In 1818, an anonymous note, generally attributed to Davy's protégé Michael Faraday, appeared in the *Quarterly Journal of Science and the Arts*; its writer pointed out that ether "produces effects similar to nitrous oxide." Ether "frolics" and nitrous-oxide parties became quite the rage. Itinerant "professors" amused audiences with demonstrations of the effects of nitrous oxide on volunteers. At such a public performance in Hartford, Connecticut, audience member Horace Wells, a dentist, noted that a volunteer who had accidentally gashed his leg appeared to feel no pain. He purchased some laughing gas and had one of his own teeth pulled out by an assistant. He felt no pain.

Wells realized that this breakthrough in pain control had a potential far beyond the field of dentistry. He asked a former partner, William Morton of Boston, to arrange a demonstration of nitrous oxide as a surgical anesthetic. Morton had been dabbling in such matters himself and had learned about ether from Professor Charles Jackson, who had been tutoring Morton privately in chemistry. He had actually tried to desensitize patients with ether, but the results had been inconsistent — now Morton was eager to see what he could learn from Wells, and his mind was on the vast amounts of money that could be made by the purveyors of anesthesia.

The demonstration was arranged at Massachusetts General Hospital, but it turned into a fiasco. The student volunteer who was to have a tooth extracted began to scream in pain: Wells, in his eagerness, had not administered enough laughing gas. Disgraced, he gave up dentistry and eventually committed suicide.

Morton, however, became even more determined to solve the problem, and he focused his attention on ether because he realized that the chief surgeon at Massachusetts General, John Collins Warren, would never agree to another nitrous-oxide demonstration. Morton did manage to convince Warren that he had a "new and improved" anesthetic to introduce. This time, the experiment was successful. A tumor was removed from a patient's jaw under ether anesthesia, and, on October 16, 1846, the era of painless surgery dawned. Within weeks, surgeons around the world were employing ether.

The public learned about ether from a most unusual source. The famed magician Jean-Eugene Robert-Houdin incorporated the new discovery into his act. He had designed a "suspension" illusion whereby his son appeared to float in the air, defying gravity. Robert-Houdin had the idea of wafting ether fumes over the audience while the illusion was being performed to create the impression that the ether vapors were actually lifting the young man. Thousands learned about the existence of ether in this fashion. The practical benefit of this stage illusion was that it alleviated people's fears about ether — remembering Robert-Houdin's stage effect, they would think the worst that could happen to them under ether anesthesia would be that they might float right off the operating table.

The years following the introduction of ether saw a bitter battle between Wells, Jackson, and Morton over who was the rightful discoverer of anesthesia. The truth is, it was none of them. The discoverer of anesthesia was most likely Crawford Long, a well-trained rural Georgia physician. When the United States Congress was trying to decide who among Wells, Jackson, and Morton should get a $100,000 award for the discovery that had so dramatically alleviated human suffering, its members received a letter from Long describing how he had used ether to remove cysts and even amputate toes at least four years before Morton's classic demonstration in Boston. Being a country doctor plying his trade outside the mainstream of academia, he had never bothered to publish the results of his ether experimentation.

Because of the confusion, the prize was never awarded. Wells committed suicide two years later, Morton died of a stroke just after one of his petitions to Congress was rejected, and Jackson ended up in an insane asylum after he chanced upon Morton's grave in a Boston cemetery and noted that the tombstone inscription declared him the "Inventor of Anesthetic Inhalation." As for Long, he outlived the others and enjoyed a long career; he died of a stroke while attending to an etherized patient who was painlessly giving birth. A statue of Crawford Long now stands in the US Capitol, a tribute to the man who made what is perhaps the greatest medical discovery of all time.

HORMONES AND THE HAIR CHALLENGED

I will always remember my high-school chemistry teacher, not because of the way he taught chemistry, but because of the way he combed his hair. The few hairs that still clung to his nearly bald head were allowed to demonstrate their full growth potential; they were slicked over the man's shiny dome in a valiant attempt to defy nature's decree.

This type of thing is frowned upon by the Bald Headed Men of America, an organization dedicated to glorifying baldness. Headquartered in — where else? — Morehead, North Carolina, the BHMA produces a barrage of slogans like, "Fight drugs, plugs, and rugs," and, "Hairiness is an outdated evolutionary idea." The association also publicizes the comments of baldness boosters such as the Harvard professor who claims that baldness is caused by great intelligence: the brains of smart men, he insists, grow larger than the brains of men displaying average or low intelligence, stretching the scalp until it is too tight to hang onto hair. The only thing being stretched here, of course, is our credibility.

"Do you really want to waste your hormones growing hair?" asks the BHMA, trying to send a shiver down the spine of any man who has considered his full head of hair a blessing. Can it possibly be that nature has compensated the hair challenged with other hormonal attributes? There is, to be sure, a connection between hormonal activity and hair growth. However, scientific evidence, to the dismay of bald men everywhere, does not support the claim that bald men are sexier. But let's start at the beginning.

In the 1930s and 1940s some researchers concluded that certain forms of mental illness that lead to aggressive behavior are initiated by an abundance of male hormones. To them, the remedy seemed simple enough: castration. The procedure was routinely carried out on troublesome patients at a Kansas mental hospital, and this caught the attention of anatomist James Hamilton of Yale University. Hamilton, who had a special interest in the effect of male hormones, was granted permission to study the castrated mental patients. One of these patients had a twin brother who came for visits. Hamilton noted that the man was completely bald and learned that he had been so for 20 years. His identical twin, the hospital inmate, had a full head of hair. Was there a connection between male hormones and hair growth? Hamilton obtained permission to inject the hairy patient with testosterone, the male hormone he had been deprived of by castration. Within six months, the man was as bald as his brother.

It seemed clear that testosterone could cause baldness. Testosterone was also known to be responsible for the sex drive. It therefore seemed logical to conclude that bald men are indeed blessed with an unexpected benefit. Alas, further studies showed that bald men do not have more circulating testosterone; rather, the substance is metabolized differently in their follicles, those little pockets in the scalp from which hair grows.

This information came to light under circumstances just as unusual as those of the mental-hospital incident. Doctors in Santo Domingo in the Dominican Republic had long been interested in the unusual number of boys they were seeing who were afflicted with a condition called "guevedoces." This translates as "penis at 12," and applies to boys in whom genital development is delayed until adolescence but then proceeds normally. Two lingering effects of this condition have been documented: as the boys grow older, their prostate glands remain unusually small and they do not become bald.

The underlying chemistry is fascinating. The condition is characterized by a deficiency in an enzyme called 5-alpha-reductase, which converts testosterone to its metabolite, dihydrotestosterone (DHT). It seems that prostate enlargement as well as male-pattern baldness are associated with the work of DHT in hair follicles and that bald men have very efficient enzymes to bring about its formation. Researchers have also noted that alcoholics are rarely bald; this would appear to make sense, since constant alcohol consumption reduces the body's ability to convert testosterone to dihydrotestosterone.

So, it came as no great surprise to scientists when they discovered that finasteride (Proscar), a drug developed to block the action of 5-alpha-reductase in order to counter benign prostate enlargement in men, had the side effect of promoting hair growth. In fact, under the name of Propecia, the drug is now available as the world's first oral antibaldness medication. The recommended dose is one-fifth that prescribed for prostate enlargement.

Propecia is no miracle — only about 15 to 20 percent of subjects show cosmetically effective hair growth. This is further tempered by the observation that about 2 percent of subjects experience sexual dysfunction, a point that is gleefully publicized by the Bald Headed Men of America. In any case, Propecia is poised to take its place alongside Rogaine (minoxidil), the only other approved remedy for male-pattern baldness. The mode of action of minoxidil, which was first introduced as an antihypertensive agent, is not known. Doctors simply discovered that patients taking the pills began to exhibit hair growth. Eventually, a topical version of the product was developed and found to yield satisfactory growth in roughly 10 percent of users, both male and female, but only as long as it was dutifully applied twice a day.

These are not great statistics, especially when we take into account that rubbing just about anything on a bald head can temporarily stimulate dormant follicles into action. Hippocrates, the father of medicine — who was, according to ancient illustrations, bald as a billiard ball — had success with an ointment made from horseradish and pigeon droppings. Others have proclaimed the benefits of Chinese herbs, onions, vitamins, placenta extracts, goose dung, cow urine, and bull semen. A highly touted "European formula" is based on polysorbate 60, a salad emulsifier.

Each of these remedies has its devotees who are convinced that they have found an answer to one of life's greatest challenges. Hope rules eternal: studies show that even when an examination of baldness-remedy users reveals that no new hair has grown, 20 percent of subjects are convinced it has.

The number of ineffective baldness remedies that have been promoted to the public over the years is positively hair-raising. Maybe it's time to give up and listen to the Bald Headed Men of America when they say, "The Lord is just and the Lord is fair; he gave some people brains and the others hair."

Going Nuts about Selenium

I think I'm going nuts. Brazil nuts. I'm going to eat a few every week. Why? Because they are one of the best sources of selenium, a mineral that is arousing a great deal of interest in the scientific community as a possible protective factor against disease, mainly cancer. With all that we hear about increasing cancer rates, incorporating more selenium into our diets may be just the kind of nutty idea we need. But let's start at the beginning.

Way back in the 1930s, Chinese authorities noted that Keshan County had an unusually high incidence of young people with a type of heart disease known as cardiomyopathy. Scientists could not find an explanation for the virtual epidemic until hair analysis offered a clue. The average concentration of selenium in the hair of people living in areas with a high rate of heart disease was less than half that of people living elsewhere. This did not mean that the disease was necessarily caused by low levels of selenium in the body, but the theory was worth investigating. Accordingly, the Chinese government decided to supplement the diets of those at risk with selenium and, remarkably, was successful in eradicating this form of juvenile cardiomyopathy.

Further research revealed that the selenium content of the soil, and hence of the crops grown in it, varied greatly in China and was particularly low in Keshan County. This prompted an investigation into the possibility that conditions other than "Keshan disease" were also linked to the selenium content of the diet. Blood from blood banks around China was analyzed for selenium, and regions were ranked according to the amount of selenium found. The results were dramatic: areas with the highest blood levels of selenium had the lowest cancer-death rate.

These results were confirmed in other areas of the world as well. In the United States, the Dakotas and Wyoming have soils rich in selenium and low cancer rates. A wide-ranging study conducted in 1977 showed that in 27 countries dietary selenium was inversely correlated with death from several types of cancer.

Such associations are intriguing, but they cannot prove cause and effect. For that, we need studies usually referred to as intervention trials. These involve two groups of subjects treated in an identical fashion except that one group is given the substance being tested.

There have been a large number of animal intervention studies demonstrating that selenium can protect against tumors. For example, rats exposed to benzopyrene, one of the cancer-causing compounds in smoke, develop fewer tumors when pretreated with dietary selenium. But rats aren't human, and that's why American researchers decided to carry out a large-scale selenium intervention study on people.

They chose subjects who had been diagnosed with skin cancer, hoping to see changes in the progress of the disease that could be linked to a daily supplement of 200 micrograms of selenium. The study, which was designed to last at least seven years, was abruptly terminated after just four and a half years. While the selenium had no effect on skin cancer, the researchers had noted 63 percent fewer cases of prostate cancer, 58 percent fewer cases of colorectal cancer, and 45 percent fewer cases of lung cancer in the selenium group. These results were so astounding that the researchers decided it would be unethical to carry on with the study without telling subjects in the placebo group about the cancer-protecting effect they had found.

Predictably, the popular press enthused over the outcomes of the study and sales of selenium supplements skyrocketed. Then the naysayers made themselves heard. The study was done in the South, they said, where the soil is low in selenium. The trial was therefore just remedying a natural deficiency, and the same results would not be seen in other areas. Furthermore, the supplement used in the study was a special yeast grown in a selenium-enriched medium and was quite different from the sodium selenite found in most supplements. These are valid arguments, but they certainly do not mandate a wholesale dismissal of the study's impressive results, especially when we consider that a Chinese intervention study came up with similar findings. Over 200 people who had been exposed to hepatitis B and were therefore at greater risk for cancer were treated daily either with 200 micrograms of selenium or a placebo. After four years, there were five cases of cancer in the placebo group and none in the selenium group. Very interesting. So is the observation that viruses reproduce more easily in a selenium-deficient host. Zaire, the country where the HIV virus first appeared, has a selenium-deficient population. Even impaired sperm motility has been linked with low levels of selenium.

The plot thickens when we discover that there is a chemical rationale for the protective effect of selenium. The enzyme glutathione peroxidase is important to the proper functioning of the body's immune system; its role is to neutralize free radicals, which can cause tissue damage. And guess what? Selenium is an integral part of this critical enzyme. There is also some evidence that selenium can cause cancer cells to die before they replicate.

But before anyone starts gorging on selenium pills, consider this: not all cancer studies have found a link with selenium. One of the largest examined the selenium content of toenail clippings from over sixty thousand nurses. It is well established that the selenium content of nails is a reflection of dietary intake, but the study found no connection between breast cancer and selenium levels in the toenails. Then there is the problem of toxicity. About 10 years ago in England, a man visited an emergency room for the third time in a month complaining of vomiting and diarrhea. When hospitalized, he improved, but he became sick again when he went home. His hair and fingernails started to fall out. His doctors went on the alert when he mentioned that his girlfriend's cooking left a bad, garlicky taste in his mouth — a classic sign of selenium poisoning. As it turned out, the woman was angry about his refusal to leave his wife and children for her. She had a friend purchase selenous acid from a hobby shop, where it was offered as bluing for gun metal. A few drops mixed into each meal was her revenge. But the fiendish plot was foiled, and the would-be poisoner got five years for her trouble.

The point is that selenium in high doses can be toxic. And these doses do not have to be much higher than the usual supplemental doses of one hundred to two hundred micrograms. At eight hundred micrograms daily, hair loss, fingernail malformation, and gastrointestinal problems have been noted.

Cows, horses, and sheep that graze on plants grown in soil that is high in selenium have been known to totter about in a clumsy fashion — farmers say that they have the "blind staggers."

We certainly don't want to be staggering blindly as we decide whether to take a selenium supplement. There seems to be enough evidence to suggest that we should strive for a total daily intake in the range of two hundred to three hundred micrograms. Fish is a good source of selenium, but, as we have seen, grains and vegetables are variable depending on soil conditions. Adding selenium to fertilizer is one way to increase intake. In fact, this is exactly what is being done in Finland, a country with notoriously low selenium levels in its soil. Garlic is especially adept at picking up selenium from the soil in which it is grown. There has even been talk of spraying tobacco plants with a selenium solution before harvest to reduce the risk of tobacco-induced cancers (of course, anyone really interested in reducing the risk can just give up smoking).

North American soil is richer in selenium than Finnish soil, so for North Americans a one-hundred-microgram dietary supplement is innocuous and may provide some nutritional insurance. The best ones contain selenium incorporated into an amino acid like methionine. Supplements are especially appropriate for men at risk for prostate cancer or for those who already have the disease. But why reach for a supplement when Brazil nuts are an ideal source? The Andes soil in which they are grown is high in selenium, and each nut contains about 120 micrograms in a readily absorbable form. It may be a tough nut to crack, but it's well worth the effort.

HYPE, HOPE, AND GINSENG

Probably what first attracted the Chinese to the odd-looking root was its shape. Its decidedly human appearance earned it the name "ginseng," or "manlike." The root's uncanny resemblance to the body undoubtedly encouraged people to ingest it, and soon the first claims about the root's beneficial effects were recorded. Ancient Chinese manuscripts speak of ginseng's ability to brighten the eyes, open the heart, invigorate the body, and prolong life. Since that time, the claims have become even more extravagant.

Proponents now suggest that ginseng can increase energy levels, improve immune function, rev up the sex drive, enhance athletic performance, boost mental ability, lower cholesterol, diminish menopausal hot flashes, alleviate insomnia, act as an anti-inflammatory agent, and reduce the risk of cancer. In light of such claims, there is little wonder that ginseng's botanical name, "panax," derives from the name of Panacea, the Greek goddess who could heal all ailments — but while ginseng is certainly intriguing as a medicinal substance, it's no panacea.

When we try to evaluate the potential of ginseng, we encounter several problems right off the bat. First of all, there are several species of ginseng. *Panax ginseng* is native to Asia, whereas *panax quinquefolius* is found in North America. Then there is Siberian ginseng (*eleutherococcus senticosus*), a more distant relative. The chemical composition of these species is quite different; in fact, there can be significant variation between two plants from the same family grown under different climactic conditions. Dozens of compounds have been isolated from each type of ginseng, and there are no standardized extraction techniques.

The best candidates for biological activity are the "ginsenosides" (also referred to as triterpenoid saponins), some of which can release steroids upon ingestion, but there is often no way of knowing to what degree these compounds are present in a commercial product. No labeling requirements exist as yet, though some manufacturers will list the concentration of ginsenosides. To complicate things further, at least 11 ginsenosides have been identified and their relative activities are unknown.

The labeling problem was illustrated dramatically when Swedish researchers examined 50 products sold in 11 countries and found that 6 samples contained no active ingredient and the concentration of ginsenosides in the other samples ranged from 2 to 9 percent. One sample, sold in the United States, contained no ginseng derivatives at all but had undeclared ephedrine, a potentially dangerous stimulant. This came to light when an athlete was accused of doping himself based upon a positive urine test for ephedrine. He realized that the only possible explanation was the ginseng preparation he had taken. An extract of *periploca sepium* (Chinese silk vine) is sometimes passed off as Siberian ginseng (and remember that Siberian ginseng isn't even ginseng). A 30-year-old Toronto nurse who was taking Siberian ginseng for irritability and mood swings during pregnancy gave birth to a baby with thick pubic hair, and more hair covered the infant's forehead. *Periploca sepium* has obvious hormonal effects.

There is some concern that ginseng itself contains estrogenic compounds that may pose a threat to people with a family history of breast cancer. In fact, vaginal bleeding has been seen in heavy ginseng users, suggesting hormonal activity. Painful breasts, skin rashes, insomnia, and diarrhea have also been reported. Ginseng may also affect glucose levels in the blood, causing problems for diabetics.

Some of the compounds found in ginseng are similar to digitalis, a drug used to treat congestive heart disease. Conceivably, then, certain ginseng preparations can also have an affect on the heart and should not be used indiscriminately by heart patients.

The possible benefits of ginseng have been well studied in the laboratory and through animal testing. There have been several interesting findings. Ginseng, for example, has been shown to improve memory in rats. Mice placed on a ginsenoside-rich diet before being exposed to colon-cancer carcinogens developed fewer tumors. As far as humans go, one Korean study suggested that ginseng users have a significantly lower incidence of cancer. Fresh ginseng extract and powder have been associated with a reduced risk of the disease but ginseng juice and tea have not. Of course, ginseng users may have had some other lifestyle effect that may account for the difference in cancer rates.

The most intriguing experiments on the effects of ginseng have involved endurance studies. Mice given ginseng extract will run for longer intervals on a treadmill and will swim for longer periods before becoming exhausted. Unfortunately, there have not been many well-controlled human studies that corroborate this effect. One study did show that middle-aged Swedish men who receive ginseng for eight weeks have a greater capacity for physical work. Russian and Japanese researchers have linked increased stamina and endurance with ginseng consumption.

In light of such findings, scientists have begun to describe ginseng as an "adaptogen," meaning that it may somehow enhance the body's ability to adapt to physical stress and, possibly, mental stress. Students have spoken of taking ginseng to reduce the feelings of stress associated with exams. While this may work, I suspect that studying is more effective.

While ginseng's potential is certainly tantalizing, the fact is that we do not yet know which of its components are the active ones. Yet even if we had this knowledge it might not be of much help because the composition of commercial preparations is, in general, a mystery. Perhaps the best way to try ginseng is to consume about two grams of fresh powdered root a day. This will not be cheap: ginseng is the most expensive legal crop in the world.

A good alternative is a standardized extract containing at least 7 percent ginsenosides. This may help us to feel more energetic, and it may actually increase our physical endurance, but keep in mind that potential cross-reactions between ginseng and other medications have not been adequately investigated and that, as a rule, skepticism is warranted if something sounds too good to be true. Remember that Panacea, the goddess who could heal every ailment, was a mythological character.

Vitamin E Gets an A from Researchers

Anytime the issue of vitamin E comes up people want to know the bottom line. Should they take supplements, and, if so, which ones and how much? Well, the bottom line is that there is no bottom line. But there sure is a lot of fascinating information.

Vitamins are substances that we must consume, in small amounts, in order to maintain good health. In the 1920s, researchers discovered that male rats lacking a fat-soluble substance in their diets became sterile, and female rats were unable to carry their young to full term. This substance was named vitamin E, or tocopherol, deriving from the Greek words tokos, or "birth" and phero, or "carry."

Chemical analysis revealed that vitamin E is actually composed of eight related compounds. These have differing abilities to prevent reproduction problems in rats, with d-alpha-tocopherol possessing the greatest biological activity. This substance is amenable to laboratory synthesis but when made in the lab it inevitably forms with its nonidentical mirror-image substance, "l-alpha-tocopherol," which does not exist in nature. The "l-isomer," as it is called, has far less biological activity than the "d."

Since the eight naturally occurring components of vitamin E and the synthetic "l" version all have different biological activities, the researchers recognized the need for a standardized unit of measure for vitamin E activity. Weight would be misleading because one milligram of synthetic vitamin E — which is composed of the active "d" and the less active "l" forms — does not have the same effect as one milligram of pure "d." Therefore, the term "international unit" (IU) was coined to represent the biological activity of one milligram of synthetic vitamin E. By this scale, d-alpha-tocopherol has an activity of 1.49 IU. This means that any tablet labeled as having two hundred IU of vitamin E will have exactly the same capacity to prevent reproductive problems in rats, although it may not have exactly the same composition as another tablet labeled two hundred IU.

"Natural" vitamin E tablets are generally made by extracting pure d-alpha-tocopherol from soybeans, while the synthetic version consists of equal amounts of d-alpha-tocopherol and l-alpha-tocopherol. Neither contains any of the other seven components, which are found in nature along with the d-alpha form.

However, the great interest these days in vitamin E has nothing to do with its effects on reproduction. Instead, that interest arises from vitamin E's ability to neutralize some of the negative effects of oxygen in the body — that is, to act as an antioxidant. A recently conducted University of California study examined this antioxidant activity of vitamin E and came up with a surprising finding: gamma-tocopherol, one of the vitamin E components not found in supplements, offers protection from some harmful oxygen by-products, such as nitrogen oxides, that are ignored by d-alpha-tocopherol.

The researchers also discovered that large supplemental doses of alpha-tocopherol have the effect of reducing the body's absorption of gamma-tocopherol from food sources. So, eventually, the composition of supplements may have to be modified for optimal results. A combination of alpha-tocopherol and gamma-tocopherol may prove to be desirable.

But let us also remember that the reason for the popularity of vitamin E is the impressive number of studies demonstrating the benefits of using the supplements that are now available. For example, two studies done by Harvard researchers involving some 135,000 health professionals found that those taking vitamin E supplements had one-third fewer heart attacks.

A British study examined the effects of giving vitamin E supplements to men who had clogged arteries as determined by angiograms. After 18 months, the men who took supplements had a 77 percent reduced risk of nonfatal heart attack. These results are probably due to vitamin E's ability to prevent the oxidation of LDL cholesterol — commonly known as "bad" cholesterol — to a form that damages arteries.

Other intriguing studies using vitamin E supplements have pointed to reduced lung damage from air pollution, a lower incidence of cataracts, enhanced immunity, a better response to hepatitis vaccines, and a slowing of the progress of Alzheimer's disease. Animal studies have indicated that the substance may offer protection against certain types of cancer, and in humans vitamin E is know to block the formation of nitrosamines, which are potent carcinogens.

To date, however, no one has done a classic intervention study — that is, two identical groups have not been treated with vitamin E or a placebo over an extended period while their health status is monitored. In the absence of such a crucial study, we are forced to make judgments concerning the advisability of supplements based on the less direct types of evidence detailed earlier. In doing so, of course, we must also take into account potential harmful effects. Luckily, these are few.

Vitamin E does have an anticoagulant, or blood thinning effect, which may, in fact, partly account for the lowered incidence of heart disease in some of the supplement studies. This effect may also indicate that vitamin E enhances the effect of other medications, such as aspirin and coumadin, that thin the blood as well. Anyone taking such medication should consult a physician about the wisdom of taking vitamin E.

There has also been some concern that vitamin E's ability to improve immune function may not be entirely desirable. In theory, enhanced immunity may worsen autoimmune diseases, such as arthritis, in which the body's immune system mistakenly attacks its own tissues. This, however, has not been observed in practice. Rarely, side effects such as nausea, diarrhea, cramps, fatigue, headache, blurred vision, and rashes have been noted.

So where does all of this leave us? Should we just wait until the definitive studies are in? Unfortunately, it's unlikely that there will ever be a study that is decisive in everyone's eyes. All we can do is make some good scientific guesses based on the hundreds of articles that have been published in scientific journals.

Distillation of the evidence suggests that a daily intake of about two to four hundred IU of vitamin E is optimal. Ideally, this should come from food that has a balance of all eight vitamin E components. But the truth is that it is very difficult to consume this much dietary vitamin E, especially given the fact that the best sources are nuts, seeds, and vegetable oils, which are all high in fat.

A two-hundred-IU-daily supplement seems like a good idea, even though the currently available pills do not contain all the vitamin E components found in food. Perhaps the most telling argument for supplementation is that while many vitamin E researchers are unwilling to make this recommendation to the public at large, most do take supplements themselves.

THE WHIFF OF ROMANCE

The scent was absolutely intoxicating. He moved towards her in the dark. First, he gently sniffed behind her ear, and then he moved down to explore her more erogenous zones. Testosterone surged through his body. He sensed she was ready. There was now no holding back, and they finally melded into one. A few weeks later, a litter of Syrian golden hamsters was born.

Thanks to some fascinating research conducted at Rockefeller University, we now know more about the love life of these rodents than of virtually any other animal. When the female comes into estrus she drags her rear end on the ground and deposits a watery secretion composed of hundreds of compounds. One of these, dimethyl disulfide, arouses the interest of the male. We refer to it as a pheromone, a chemical signal transmitted between members of the same species to elicit a specific activity. With great enthusiasm, the male seeks out the source of this pheromone and tracks down the female. Interestingly enough, dimethyl disulfide just attracts the male, it doesn't excite him sexually. That job is left to another pheromone — a nonvolatile protein.

Once the hamster has been attracted by the dimethyl disulfide, he begins to sniff and lick the female, eventually focusing on her reproductive region. Now the action gets hot and heavy. There's lots of squirming and smelling until she finally assumes the mounting position. The male's appetite has been completely whetted by the proteinous secretion, and he wastes no time taking advantage of the presented opportunity. He intromits and then, typically, runs away.

The protein that elicits all the furious activity has actually been isolated through some painstaking chemical work and appropriately named aphrodisin. How did the researchers prove that it really is that substance which triggers copulation? By using an anesthetized male stand-in. They propped up this surrogate in a mating position and anointed him with dimethyl disulfide to pique the curiosity of another male. Indeed, the unsuspecting rodent approached, sniffed, and licked a little, but he soon became uninterested. However, when the surrogate's hindquarters were treated with aphrodisin, the test male's licking became more and more vigorous until he eventually mounted the "female" and attempted to satisfy his urge. The researchers in attendance were no doubt gratified to observe this bizarre little scene: it proved to them that aphrodisin is indeed a copulatory pheromone. One suspects that neither male hamster shared their elation.

You may think aphrodisin is pretty powerful stuff, but it's nothing compared to periplanone-B, the pheromone of the American cockroach. Males will literally break their legs trying to get to the source of this compound. Interest in cockroach sex attractants arose when researchers at the US Army's Natick Laboratory began to wonder how these tiny creatures find each other in the dark (perhaps with some vague notion about enlisting the roaches for military action). It soon became apparent that some volatile chemical-messenger substance was involved, because the males in their little cages began to jump frantically about when virgin females were brought into the lab in a box. The army scientists were, however, unable to isolate the chemical that had so excited the male cockroaches. It took the CIA to do that.

At the height of the Cold War, the Central Intelligence

Agency hatched the idea of using the cockroach scent to track Soviet spies. CIA scientists concentrated substances collected by passing air over a milk jug filled with ten thousand virgin females. How they actually determined the sexual status of the bugs is, of course, a state secret, but after nine months they did isolate nine milligrams of the pheromone, which they then dissolved in a solvent to make a "cockroach perfume." The idea was to apply a few drops of this concoction to a person under surveillance and to track that person with a cage containing virile male roaches that would stampede at the whiff of the pheromone. The agent equipped with the cockroach detector could remain safely out of sight of his quarry because those male roaches are so sensitive to the smell.

Whether this ingenious scheme was ever actually implemented is not clear, but we do know that it has taken about 30 years to identify the exact molecular structure of the compound that has come to be known as periplanone-B. The major difficulty was just getting enough of the substance to work with, but eventually about two hundred micrograms were collected by sacrificing 75,000 female roaches. The compound proved to be extremely active. When male roaches in a jar were exposed to one-trillionth of a gram, they went into an absolute frenzy; soon, legs were broken and bodies were tattered. The passionate male roaches attempted to copulate with any three-dimensional object, including each other.

When this cockroach aphrodisiac was isolated and identified, the question became what to do with it. The world had no need for increased amorous adventures among cockroaches. In fact, while in most people's minds anything that induced roaches to beget other little roaches was decidedly unwelcome, if a substance could somehow be used to curb the roach population, that was a different story. An application towards this end was obvious: why not use periplanone-B to trap unwelcome guests in the home? As it turned out, the compound could easily be synthesized in the lab — availability was not a problem.

The original idea was to use periplanone-B to attract male roaches to some sort of poison inside a trap. The troublesome insects would check into the Roach Motel, but they would never check out. There was, however, a problem. As the roaches saw their fellows die around the bait, they came to associate it with death. They were soon scampering away unharmed. The answer was to use a poison that does not kill instantly. Traps now have periplanone-B, which attracts the roaches, and amidinohydrazone, which the insects dine on once they're inside. This compound acts as a stomach poison over a period of 24 to 48 hours, and the roaches never learn to connect the traps with their eventual demise.

Given that such fascinating pheromonal activity occurs among such simple creatures, might there not be some sort of parallel in humans? After all, we are also a part of the animal kingdom. We do have a special affection for fragrances derived from the anal-gland secretions of the Ethiopian civet cat or the rutting Himalayan musk deer. In a concentrated form, these secretions — known as "musk" — have an extremely foul smell, but in dilute form most humans find them very appealing. Musk is actually the most valuable animal product in the world, selling for about $44,000 a kilo, roughly four times the price of gold. Luckily, adequate synthetic analogues are readily available today.

Musk does seem to stir the human passion. Why? Napoleon, of all people, may have given us an important clue. Before returning from his battles, the emperor would write to his wife, Josephine, begging her not to bathe — her ripe fragrance turned him on. Only when she discovered the allure of a musk-based perfume did Napoleon finally stop issuing his peculiar requests: the scent of musk satisfied his primal urge the same way Josephine's sweat did. It's interesting that a compound found in human sweat, androstenol, believed by some researchers to be a human pheromone, has a decidedly musky smell — so maybe the reason we find the smell of musk so attractive is that it triggers the same receptors in our olfactory system as a human pheromone, if indeed such a thing exists. And it may.

The compound that some have dubbed a "human pheromone" was actually first isolated from swine testes in the early 1970s. Androstenol was the scent that prompted female pigs to assume the appropriate position for copulation. The effect was so dramatic that androstenol was soon made available to farmers in a spray can to facilitate artificial insemination. A tiny whiff of the substance and a sow would become ready to accept the inseminating rod calmly.

The pheromone saga then took an unexpected turn. Androstenol was found to be present in human underarm secretions. What was a pig sex attractant doing in human sweat? Had researchers accidentally stumbled upon a substance that was a human sex attractant as well?

This is not as outlandish as it sounds. Think about why Elizabethan lovers placed apples in their armpits until they were saturated with sweat before offering these "love apples" to their sweethearts. And why does a dancing woman twirl below her partner's armpit? Why do artists so often portray women with raised arms? Doesn't the Venus de Milo seem to invite us directly to sniff her armpit? Why is it that when women are asked to strike a sexy pose, they automatically raise their arm and place a hand behind their necks? Could there be something magical in armpit fragrance?

Researchers began to sweat over the problem. In an English theater, seats sprayed with androstenol seemed to attract more women. Of course, rose fragrance might have done the same.

When men were unknowingly anointed with androstenol, they were judged more attractive by women but were downgraded by men. Intriguing observations, but not the stuff of hard science.

Hard science did finally break through in the 1980s with a series of remarkable publications by George Preti of the Monell Chemical Senses Center in Philadelphia and Winnifred Cutler, then of the University of Pennsylvania. These scientists showed that when an extract of male armpit secretions was applied to the upper lips of female volunteers those women had more regular menstrual cycles. Furthermore, when female armpit extract was applied in the same fashion, the women began to have synchronized periods. There certainly seemed to be some sort of chemical communication going on.

Cutler, at that point founder and head of the Athena Institute for Women, then went one step further. She studied the sexual behavior of young women who had "female essence" derived from armpit secretions applied under their noses three times a week for about three months. These women engaged in significantly more frequent sexual activity than control subjects treated with a placebo. The female essence somehow either increased desire or made the wearer more attractive to men.

Then science took a leap of faith. Dr. Cutler began selling "Pheromone 10:13," a synthetic version of the supposed active ingredient in female essence. The actual composition of the product has not been revealed, and the only evidence of efficacy comes from anecdotal customer testimonials: "Jan from Arizona" tells us that "men have been flying around me like honeybees," while "Thelma from New Jersey" has had to rename her motor home the Love Shack.

Cutler is on firmer scientific ground with her "Athena Pheromone 10X tm" for men, described as a perfume and aftershave additive designed to increase sex appeal. The formulation of this product is also a closely guarded secret, but I have a feeling that it might also prove to be of interest to sows. In any case, Dr. Cutler has just published a peer-reviewed scientific article in which she maintains that men who use the product increase the frequency of their romantic encounters, particularly sexual intercourse. Exciting stuff.

David Berliner, former professor of anatomy and founder of the EROX corporation, has taken another approach. Back in the 1960s he noted that a skin extract he had prepared had the effect of putting his lab workers in uncharacteristic good moods. He stored this curious information in the deep recesses of his mind. Then, a couple of decades later, Berliner's interest was rekindled when scientists confirmed the presence in humans of a nasal detection system known as the vomeronasal organ (VNO), heretofore found only in animals.

This tiny slit located on the septum about half an inch inside the nostril detects molecules that have no apparent smell. It is the VNO that detects pheromones in rats, mice, and hamsters. When it is surgically removed, the animals do not respond to their mates. Here, Berliner realized, was a potential sensor for human pheromones.

Berliner worked out a way to insert a tiny probe into the nose to measure any electrical response in the VNO as a result of exposure to potential pheromones. He soon found that two compounds from his original skin extract triggered very different responses. Androstadienone caused VNO activity only in men and estratetraenol did so only in women. Although he has not been able to document any behavioral effects due to these compounds, he has built his Realm product line around them. Realm products make no claim for aphrodisiac effects: indeed, the male version contains the chemical sensed only by men and the female version the one sensed only by women. The object (so far scientifically unsupported) is to alter the wearer's mood by exposing him or her to the pheromone released by a member of the opposite sex and perhaps create an air of confidence leading to romantic interludes.

Until we have more scientific evidence about human pheromones we should perhaps just stick to musk. Maybe the advertising for musk products is not all hype. Wild Musk Oil claims to have "an exciting and provocative fragrance that releases your sensuality and does wonders for your chemistry." Maybe it does — a little dab behind the ear . . . a whiff . . . who knows?

Van Gogh's Brain

On December 23, 1888, a prostitute in the town of Arles in southern France received a strange Christmas present: a package, wrapped in newspaper, containing part of a human ear — the ear of Vincent Van Gogh. Why the artist mutilated himself in this fashion after returning home from a visit to the brothel has been the subject of intense conjecture.

Van Gogh's artistic career began in Holland. His early paintings, such as *The Potato Eaters*, were dark and somber, clearly conveying the artist's sympathy with the hard life of the poor Dutch peasant. A move to Paris, and later to southern France, had a dramatic effect on Van Gogh's style. Darkness abated and a bright light flooded his work. *Sunflowers*, perhaps the best-known painting of this period, is a veritable explosion of brilliant yellows.

Some have suggested that there was more to this light and color change than the mood-boosting effect of the sun in southern France. Van Gogh was known to suffer from psychotic fits, which at the time were identified as stemming from epilepsy. Treatment often involved digitalis, the heart-disease medication discovered by Englishman William Withering. We

believe that Van Gogh was treated with this inappropriate medication because he painted a portrait of his physician, Dr. Gachet, holding a stalk of foxglove, the plant from which digitalis is extracted.

Large doses of digitalis can cause vomiting, giddiness, and visual disturbances. Van Gogh's later paintings reflect his obsession with the color yellow, as best exemplified by *Sunflowers*. He even painted his house at Auvers yellow. Abnormalities in color perception — particularly the appearance of yellow halos around objects — have indeed been associated with the use of digitalis. Ascribing Van Gogh's medical problems to digitalis poisoning is interesting, although somewhat fanciful.

A middle-ear condition known as Ménière's disease may also have been a source of Van Gogh's problems. A review of 796 letters he wrote to his brother Theo give rise to a picture of a man suffering from attacks of disabling, recurrent vertigo and nausea, as well as sensory hallucinations. These attacks were also characterized by an inability to tolerate motion and loud sounds. Between these "fits," as Van Gogh himself called them, there were long symptom-free periods, as is often the case with Ménière's disease. Auditory hallucinations, or ringing in the ears (tinnitus), are also common. In fact, some patients speak of "cutting off their ear" or "poking a hole in it with an ice pick" to relieve the torment. The symptoms of Ménière's disease do appear to be consistent with those suffered by Van Gogh, and they present a plausible rationale for his self-mutilation on that fateful day in 1888.

There is yet another possible explanation for Van Gogh's fits of bizarre behavior, and it is based on his well-known addiction to a popular beverage of the time, absinthe, the standard ingredients of which are alcohol, oil of wormwood, anise, fennel, juniper, and nutmeg. The oil of wormwood is of particular interest here. The active ingredient in wormwood is thujone, a compound that can induce excitement of the nervous system followed by unconsciousness and convulsions. In fact, the condition induced by thujone has been studied as a model for epilepsy. Because absinthe caused so much misery, the French government banned it in the early part of this century. Its descendent, Pernod, which does not contain thujone, is still available and still intrigues people by changing (as does absinthe) from green to white when water is added to it. Compounds that are soluble in alcohol but not in water, now precipitate out.

There is no question that Van Gogh indulged heavily in absinthe during his days in southern France. When Paul Gauguin came to visit, the two painters would go on absinthe binges together and Vincent would invariably end up in a brothel (probably claiming that absinthe makes the tart grow fonder). On one of these eventful nights, he quarreled with Gauguin, threw a glass of absinthe at him, and threatened him with a razor. Then, in an agony of guilt, he retreated to his room and performed the celebrated surgery.

We will never know whether Vincent Van Gogh's suicide in 1890 was due to epilepsy, Ménière's disease, thujone poisoning, or some other factor. Some art historians claim that he sustained a cerebral lesion during birth and suggest that this was the main trigger of his aberrant behavior. In any case, the reason he shot himself behind a manure heap near the very wheat field he had depicted in his last canvas will remain a mystery. The painting is of a road that ends abruptly in the middle of a field of wild yellow wheat; according to some, the road symbolizes Van Gogh's short life.

After Van Gogh's death, Dr. Gachet planted a tree on his former patient's grave. Brother Theo died soon after and was buried elsewhere, but 23 years later relatives decided that the brothers, who had been extremely close in life, should be near each other in death, and they arranged to bury the two in the same plot. When Vincent was dug up, his casket was found to be completely entwined in the roots of the tree planted by Dr. Gachet. The doctor had unknowingly chosen a wormwood tree, the source of thujone: in death, as in life, Vincent van Gogh found himself in the clutches of thujone.

AROUND THE HOUSE

SORTING OUT THE SUDS

Periodically, I like to take a stroll down the shampoo aisle in the pharmacy to see what the current crop of products promises. Will this one or that one "energize" my hair? Add luster or remove buildup? Nourish or rejuvenate? What's the newest miracle ingredient? Sperm extract or quinine? Alpha hydroxy fruit complex or beta-carotene? It's all very confusing. Indeed sometimes it's hard to tell if you're in a pharmacy or a supermarket, with shampoos loudly proclaiming that they contain mango, papaya, apple pectin, wheat germ, or Swiss vanilla, whatever that may be. It all sounds outrageous. Revlon apparently agrees. The cosmetics company has come out with a line of shampoos appropriately called "Outrageous!"

This bewildering cacophony of claims and esoteric ingredients is a product of the basic dilemma faced by shampoo manufacturers. How do you convert an essentially simple product that cleans hair into a magical lotion that increases sex appeal? The answer lies in some basic science and a lot of clever marketing.

First of all, shampoos do not feed, resuscitate, enliven, or revive hair. They can't, for the simple reason that hair is not alive. So what can shampoos do? They can clean hair. And any of them can perform the task very well because it's not difficult. All they must do is remove the thin layer of oily material, called sebum, that coats and protects the hair, because sebum also acts as a magnet for dirt and for residue from hair-treatment products.

Detergents can remove sebum very effectively and are the prime ingredients in all shampoos. They reduce the surface tension of water, making it flow more freely. Water mixed with a dissolved detergent will not form beads but will spread easily, wetting every nook and cranny of a surface.

Detergent molecules also form strong ties between oil and water. One end of the molecule is fat soluble and anchors itself in any oily residue. The other end is water soluble. As a consequence, when a detergent solution is rinsed away, fatty substances, such as sebum, are pulled away from the surface to which they are attached. The cheapest shampoo will clean hair as well as the most expensive. In fact, so will dishwashing detergent, which contains sodium lauryl sulfate, the active ingredient in most shampoos.

Sodium lauryl sulfate is a great cleaning agent, but it strips sebum from the hair so thoroughly that it leaves the hair very dry. It may occasionally cause skin irritation, but there is absolutely no truth to the rumor that it causes cancer. To prevent dryness, it is usually combined with other less harsh, but generally more expensive detergents like ammonium laureth sulfate. Quaternary ammonium compounds, known as quats, are also added to shampoos to allow for easy detangling when combing. These are the same compounds found in fabric softeners, and they work by clinging to the surface of the hair, producing a smooth finish. Foam boosters, such as cocamide monoethanolamine, are often incorporated; not that this has anything to do with cleaning — it doesn't. Hair can be cleaned effectively without any suds, but suds sell shampoos. Some shampoos also include panthenol, a molecule that can diffuse into the hair shaft and bind to proteins, strengthening their structure. Various proteins, such as elastin and collagen, or synthetic polymers, are often added with the idea that they will bind to the surface of the hair to enhance thickness. Results are variable because most of these substances end up being rinsed away with the detergent.

All shampoos, no matter how gentle, must cope with a common problem. As the protective sebum is stripped away, the outside layer of the hair, the cuticle, is exposed. In healthy hair, the cuticle consists of translucent cells that overlap like shingles on a roof. In damaged hair, these shingles are more open and ragged. As the rough adjacent hairs rub against each other, a static electrical charge can be produced by the transfer of electrons. The result is the dreaded affliction referred to as "fly-away hair."

Ideally, a shampoo will smooth down the cuticle and give it a clean coating of a sebum-like material. The smoothing effect is readily achieved by controlling the acidity of the shampoo. As long as the pH is between five and eight, the "shingles" will seal up, so all shampoos, whether they make the claim or not, are "pH balanced." The proper pH range is maintained by the addition of buffering agents such as citric acid. Humectants, which help hold in moisture, are also added. Examples are glycerol or propylene glycol, which form strong bonds to water and prevent it from evaporating.

The replacement of the soiled sebum with a clean protective coating is a more formidable task. After all, how is a shampoo to know that it is supposed to remove one oily substance and leave behind another? Modern "two-in-one" formulations with shampoo and conditioner combined have gone a long way towards solving this problem. Silicones, such as dimethicone, are relatively nongreasy materials that resemble sebum. They can coat the hair, add gloss, and provide a smooth surface to ease combing. They can also fill in the damaged areas where the cuticle has been worn away and change the reflective properties of hair, producing more shine.

Through some clever chemistry, techniques have been developed to hold silicones in suspension in a shampoo until the shampoo is rinsed away with lots of water. So, during the shampooing process, while the detergent is active, the silicones are held in a sort of state of suspended animation. As the shampoo is washed out, the silicones are activated and they coat the hair. The results can be very satisfying, but two-in-ones still can't match the effectiveness of shampooing and conditioning separately.

The search for novel ingredients that can be hyped by advertisers seems never ending. How about sperm shampoo? This pricey product touts the wonders of hyaluronic acid as the answer to bad hair days. Why? Because this acid is what sperm uses to penetrate the egg it is somehow supposed to penetrate the hair shaft. Hyaluronic acid is a good moisturizer, but the penetration stuff is a lot of bunk.

It's likely that after all of this most of you know more than you ever cared to know about shampoos. Truth be told, while there is a lot of interesting science behind these products, it all comes down to finding one you like by trial and error. Remember that those young ladies who toss their luxuriant tresses in slow motion on TV have just spent hours with expert hairdressers. So forget the ads, think of fruit salad as a dessert, take your vitamins in pill form, use sperm more appropriately, ignore the price, and find a product you like. In fact, I often wash my hair with diluted dishwashing detergent. Although it is somewhat drying, it cleans well and smells nice — but I do sometimes have an unusual desire to go and scrub a few pots and pans.

A Solution to Skunk Pollution

I remember the first time I ever smelled a skunk. I thought someone had let off a stink bomb. You see, even back then I was a lot more familiar with emissions from test tubes than from animals. Skunk secretion certainly smelled like a mixture of sodium sulfide and an acid. Such a concoction releases hydrogen sulfide, which accounts for the classic reek of rotten eggs and stink bombs, a smell potent enough to drive away any living creature, and quickly. Which, of course, is exactly what the skunk has in mind when it lets loose from the little scent glands situated on either side of its rectum.

Scientists have long been intrigued by the chemical composition of skunk aroma. Way back in 1862, the famous German chemist Friedrich Wohler received a gift of "Nordamerikan-ischen Stinkthiers" fluid from a "freunde in Neuyork." The stuff was too smelly for the great man to work with, so he gave it to one of his underlings, identified only as Dr. Swarts of Ghent. Swarts carried out the first analysis of skunk secretion and found it to be a complex mixture of many substances that distilled at different temperatures. He was able to determine, however, that the element sulfur was prevalent in the mixture, making up some 16 percent of its weight. There was a price to pay for this enlightenment: Wohler claimed that his assistant's health had been adversely affected.

Although chemists have been working on the problem of the exact composition of skunk secretions for over a hundred years, only recently have the specific smelly compounds been identified. This type of research is fraught with difficulty. First of all, how does one procure a sample? Very carefully. Skunks are trapped and anesthetized with ether. A blunt needle is then inserted into the anal sac of the animal and the contents removed by means of a syringe. This sample is subjected to analysis through an instrumental technique known as gas chromatography-mass spectrometry, which separates and identifies the components of a mixture. Literally dozens of compounds have been found in skunk extract, with seven having particularly disturbing smells. Trans-2-butene-1-thiol is the major culprit.

Now that we know this, what do we do with the knowledge? While skunk research may be academically fascinating, what we really want is a solution to the problem of the inquisitive dog or cat that has learned a lesson the hard way about the consequences of skunk chasing. How can trans-2-butene-1-thiol and its chemical cousins be neutralized? Tomato juice won't do it — that's just a myth. The only thing tomato juice will do is create a mess, leaving us with the added problem of removing tomato juice from clothing, floors, and walls. It will also turn white dogs pink.

But despair not: there is a solution, thanks to the creators of the Indiglo watch. The faces of these watches are treated with an electroluminescent material that glows in the dark. An unfortunate by-product of the manufacturing process used to make this luminescent substance is hydrogen sulfide. Not only does this compound smell awful, but it is also poisonous. A materials engineer, Paul Krebaum, working at the plant where the electroluminescent materials were manufactured, developed a technique for eliminating the smell. He designed a system whereby the air was circulated through a solution of concentrated hydrogen peroxide and sodium hydroxide. His idea was based on some interesting chemistry. Krebaum knew that sulfur binds quite readily to oxygen and that these oxidized derivatives are far less likely to smell. Experiments showed that an alkaline solution of hydrogen peroxide readily oxidizes hydrogen sulfide to odor-free sulfate. The problem of the hydrogen sulfide smell in the plant was solved.

One day, a colleague of Krebaum's came to work with a woeful tale of an encounter between his dog and a skunk. Krebaum had never considered the skunk problem before, but he knew that skunk secretions contained thiols. These resembled hydrogen sulfide chemically and should also be oxidized with his reagent. Krebaum knew, however, that he couldn't expose animals to 30 percent hydrogen peroxide — it's far too dangerous a substance, as is sodium hydroxide. The formula had to be modified. A little experimentation revealed that 3 percent peroxide would work and that the sodium hydroxide could be replaced with baking soda. Finally, a squirt of dishwashing detergent would help lift the skunk fragrance from the fur.

Here is the magic formula: take one liter of 3 percent hydrogen peroxide (available at most pharmacies), add one-quarter cup of baking soda and one teaspoon of liquid dishwashing detergent, wash the cat or dog (or child) with this mixture, and rinse with lots of water. Presto! Skunk smell is almost completely eliminated.

The latter point is an important one. People who have struggled with tomato juice and have succeeded in reducing skunk smell (not due to a chemical effect but because they have managed physically to rinse away some of the odiferous compounds) often note that the scent comes back. This is because the skunk secretions also contain compounds called thioacetates, which are not particularly smelly but over time react with moisture to form thiols. As the concentration of thiols increases, the skunk aroma returns, but under the mildly alkaline conditions created by the hydrogen peroxide recipe, these thioacetates are immediately converted to thiols, which in turn are oxidized. Therefore, the lingering smell is greatly reduced.

Most researchers are interested in eliminating the skunk stench — but not all. Skunk smell is known to keep bears away and to mask the scent of humans. This is of great interest to hunters, because their scent can often drive their prey away. Of course, nobody would want to carry around bottles of skunk extract, even if such a thing were available. The risk of an inadvertent spill would be just too great. But a clever inventor has come up with not one but two solutions to this problem. "Skunk Skreen" comes in two small bottles. One contains a thiol precursor, which forms the stinky compound when it reacts with the alkaline solution contained in the other bottle. When circumstances dictate, moisten a cloth with a few drops from each bottle, and prepare yourself for a powerful skunk-like stench. Bear beware!

As we know, the stench can also keep humans away, which is what an Alaskan inventor was banking on when he patented a "personal protector" based on skunk smell. Because a person hasn't the time to start combining chemicals when accosted by an attacker, the inventor devised a way to incorporate skunk-extract capsules in a plastic card resembling a credit card. In an emergency, all you have to do is point the card at your attacker and bend it, thereby squirting out a stream of stinking liquid. The card is smooth on one side and rough on the other to avoid accidental self-spraying.

Sounds good. Presumably, the police would have little trouble tracking the culprit, as the stench would linger for weeks — that is, unless said culprit knows enough to mix hydrogen peroxide with baking soda.

COMING CLEAN ON LAUNDRY PRODUCTS

I never know what the mail will bring. One day, it might be a new cleaning agent to try — another day, a new nutritional supplement to evaluate or a magnetic shoe insert to test for its "energy enhancing" potential. But one of the most intriguing items I have ever been asked to comment on has been the apparently magical "Laundry Disk," which claims to clean clothes without the help of detergent.

Now, I'm all for new technological developments and I certainly wouldn't mind cutting down on my detergent use for various environmental reasons, but in this case I must admit that I was skeptical from the start. I had been furnished with three colored plastic disks that rattled when shaken. A glimpse inside the perforated top and bottom revealed a number of tiny, hard beads. According to the instructions, the disks were to be placed inside the washing machine, where they would perform their wizardry for at least five hundred loads.

The "technical" information provided with the product was most interesting, to say the least. The beads, I was told, were made of "activated ceramics," which "emit far infrared electromagnetic waves that cause water molecule clusters to separate, allowing much smaller individual water molecules to penetrate into the innermost part of the fabric and remove dirt." This sounds very impressive, but infrared waves are just technical jargon for heat. Agitating the little balls in the washing machine may create a little friction, but the amount of heat this would yield is insignificant. Furthermore, the idea that this process in some way activates water molecules is nonsense.

Another claim this product makes is even more preposterous: "The activated ceramics produce an abundance of hydroxyl ions, making water molecules smaller and enhancing their solvent ability." Altering the size of water molecules is an impossibility. The product also claims to be capable of producing "ionized oxygen that kills bacteria without chemicals." Wow! These Laundry Disks must be amazing. They can kill bacteria. I wonder why we bother with antibiotics?

No list of ingredients was included with the product, but I

was eventually able to track down a "Material Safety Data Sheet," which shed some light on the composition of the Laundry Disks. The only ingredient of interest were "zoolites"; I had to assume the manufacturers were not referring to animal excreta but meant "zeolites." These fascinating mineral complexes do have the ability to soften water by removing certain metallic ions from the solution. Calcium and magnesium ions would interfere with the activity of a detergent, and for that reason zeolites are added to commercial detergents, but it is hard to imagine how they could produce any significant effect in the case of the Laundry Disks.

Of course, the main issue is not whether the effectiveness of the product can be justified on theoretical grounds but whether the product itself actually works on dirty laundry, so I enlisted my wife's help with the appropriate experiments. We washed, separately, colored and white laundry using either detergent, Laundry Disks, or just plain water. The results came out in the wash. The detergent worked best; the clothes washed using the Laundry Disks and the plain water were indistinguishable.

Amazingly, the plain water actually did quite a good job — people tend to underestimate the ability of water to remove dirt, and this probably accounts for the glowing comments from satisfied customers featured in the Laundry Disk brochure. It is also likely that small amounts of detergent left in clothes from previous washes contributes to the perception that the Laundry Disks are effective cleaners. In general, laundry can be cleaned very effectively with less detergent than manufacturers recommend.

Now that we've gotten the nonsense out of the way, let's examine the real chemistry of cleaning clothes. What's in detergents, anyway? The main ingredient, which plays a dual role, is a surfactant, an agent that increases the wetting ability of water by allowing it to spread more freely into the minute crevices of the item being washed. When a drop of water is applied to a surface it normally forms a bead due to a particular property of water — surface tension. This simply means that the molecules at the surface have a very strong attraction for each other and do not want to let go.

The molecules of a surfactant force themselves between the water molecules and reduce the surface tension, allowing the water to "wet" an article more thoroughly. Perhaps it is this phenomenon that the copywriters for Laundry Disks have tried to capitalize upon. The surfactant molecules also help in removing grease. One end of the molecule has an attraction for grease, while the other has an affinity for water. When the grease is thus anchored to the water, it, as well as any dirt imbedded therein, can be removed by flushing with water.

Surfactants can be inactivated by minerals dissolved in water. In the case of soap, which is also a surfactant, the interaction leads to the precipitation of soap from solution — the end result is the classic bathtub ring. The surfactants used in detergents do not precipitate out of solution, but they do lose their efficiency in the presence of dissolved calcium or magnesium. Accordingly, water softeners such as phosphates, aluminosilicates (these are the zeolites), or good old-fashioned sodium carbonate (also known as washing soda) are added to bind the troublesome minerals and allow the surfactant to carry out its job.

There is more. If water gets into the detergent box, the product will clump together and form large, cumbersome chunks. To ameliorate this problem, sodium sulfate, which has the ability to absorb moisture, is added. Other ingredients may include an enzyme that breaks down protein stains, such as blood, or chemicals that absorb ultraviolet light and convert it to visible light for that "whiter than white" effect. Perfumes are often thrown into the mix, although there is less of a tendency to do so these days because of the slim possibility of allergic reactions.

Detergent manufacturers even maintain our washing machines: they add sodium silicate to their preparations to prevent corrosion. This substance reacts with metal to form a thin, water-impervious protective layer, ensuring that our washing machines keep functioning so that we will keep buying more detergent. Still, isn't all this better than spending eighty dollars for three ineffective Laundry Disks and a lot of pseudo-scientific lingo? I did eventually find a use for my disks — they made great baby rattles. Maybe those "far infrared waves" have a calming effect.

THE LITTLE MERMAID AND OPTICAL BRIGHTENERS

I always enjoy spending a few days at Disneyworld in Florida — lots of heat, lots of people, and lots of fun. But who would have thought that the Adventures of the Little Mermaid attraction at MGM Studios would turn out to be a great chemical experience? Imagine this. You enter a dark, mercifully air-conditioned theater built to resemble an underwater cave. Mist fills the air, the tiny water droplets sparkling in shafts of spectacular laser light. There are audible gasps all around as an underwater seascape, complete with colorful fish, plants, and mermaids, unfolds. Fish swim, octopuses float, and various other fantastical creatures frolic happily. There is no sign of hands or support cables. The effect is truly magical. How is it all done? By exploiting the right chemistry, of course.

Audience excitement at this attraction is actually produced by excited molecules onstage. Some substances possess the ability to absorb light of one type and re-emit it in a different

form. In scientific terms, the light that is emitted is of a different wavelength from the light that was absorbed. A typical
example would be the fluorescence of certain fabrics under the
black light of a disco.

"Black light" is really ultraviolet light and is therefore invisible to the human eye. Fluorescent molecules can convert
ultraviolet light to visible light, thereby giving an object the
appearance of "glowing in the dark." The marvelous Little
Mermaid stage effect is produced by using a variety of fluorescent paints and by bathing the stage in invisible ultraviolet
light. The performers and the support structures that create the
illusion are obscured by substances that do not fluoresce. The
effect is truly spectacular.

Fluorescence is employed in other areas of life as well. We've
often seen ads that proclaim a certain detergent will make our
laundry "whiter than white," and, in a sense, this really can be
done. Fluorescent materials called optical brighteners are
incorporated into the detergent formulation and adhere to fabrics just as dyes do. They transform the invisible ultraviolet
portion of sunlight into visible light, creating brightness; laundry that is quite dirty can still appear to be brilliantly clean.

Even the packaging of cleaning agents is printed with fluorescent dyes to catch the shopper's eye. Just walk into a disco
with a package of Cheer under one arm and a package of Tide
under the other for a dramatic display and some good clean
fun.

We can shed some further light on fluorescence by considering the omnipresent fluorescent light. Just how does it work?
Fluorescent tubes contain a small amount of mercury vapor.
The application of an electrical current causes a stream of electrons to traverse the tube, and these collide with the mercury
atoms, which become energized and consequently emit ultraviolet light. The inside of the tube is coated with a fluorescent

material, such as calcium chlorophosphate, which converts the invisible ultraviolet light into visible light. The same idea is behind color television pictures. The TV screen is coated with tiny dots of substances that fluoresce in different colors when they are excited by the beam of electrons that scans the picture.

Fluorescent materials had practical applications even before we dreamed of color television. One of the most amazing of all fluorescent materials is a synthetic compound appropriately called fluorescein. Under ultraviolet light, it produces an intense yellow-green fluorescence, which, during World War II, was responsible for saving the lives of many downed flyers. Over a million pounds of the stuff were manufactured in 1943 and distributed to airmen in little packets to use as sea markers. Since the fluorescence is so potent that it can be seen when the concentration of fluorescein is as little as 25 parts per billion, rescue planes easily spotted the men in the ocean. Extensive use of fluorescein was also made on aircraft carriers. The signalmen on deck wore clothes and waved flags treated with the compound, which was then made to glow by illumination with ultraviolet light. As the incoming pilots could spot them clearly, the need to use runway lights, which would have drawn the attention of enemy aircraft, was eliminated.

Certain natural substances also fluoresce under ultraviolet light — urine, quinine, and moose fur are interesting examples. Prisoners have been known to exploit this property of urine by using their urine as an invisible ink. Tonic water, which contains quinine, will also fluoresce mystically. What about the moose fur? Well, in Canada, the US, and Sweden there are hundreds of accidents each year involving automobiles and moose. Some of these collisions result in fatalities. Certain car manufacturers are now considering fitting their vehicles with UV-emitting headlights to reduce the number of moose collisions.

As I sat watching the fluorescent frolics of the Little Mermaid, I was reminded of an intriguing ultraviolet episode of a bygone era. In the 1970s, some commercial laundries hatched a clever scheme to help them identify and sort laundry: they would mark the clothing with ink that was invisible under ordinary light but that would fluoresce under ultraviolet light. To sort the laundry, workers would just shine UV lamps on the fabric and the markings would become immediately visible — no need for tags or messy laundry pens. It seemed like a great idea, and it was. That is, until black-light discos became popular. When laundry numbers fluoresced eerily on his shirt as he approached the dance floor, I don't imagine the average club patron was appreciating the interesting chemistry involved.

THE WRONG CHEMISTRY

Chemistry has always been associated with the act of mixing things together. Actually, that's how the science got its start: our distant ancestors mixed starch with yeast and made alcohol; they mixed animal fat with wood ashes and made soap; they mixed sulfur, charcoal, and saltpeter and made gunpowder. These useful commodities whetted their appetites, and they persevered, mixing substances with the hope of producing other handy things. The results weren't always satisfactory. The Egyptian's attempts to cure blindness by pouring a mixture of pig eyes, antimony, rust, and honey into a sufferer's ear did not work. Neither did Hippocrates's endeavor to cure baldness with a mixture of opium, horseradish, and pigeon dung.

Modern chemical experimentation has shown us how to mix silver, tin, and mercury to fill cavities in our teeth; how to mix baking powder with flour to make cookies; and how to combine ethylene glycol with terephthalic acid to make polyester.

We can even mix nucleotides together to make DNA. But, of course, we've also learned that there are some substances that should *never* be combined.

A couple of years ago, a lady complained to a neighbor about an infestation of mice in her house. The well-meaning neighbor offered this suggestion: mix some toilet-bowl cleaner with bleach in a container and leave the concoction in the house overnight; it's guaranteed to get rid of the mice. What she neglected to say was that it would likely get rid of the human inhabitants as well. Permanently.

Chemically speaking, bleach is a solution of sodium or calcium hypochlorite. When mixed with any acid, it releases highly toxic chlorine gas. Most toilet-bowl cleaners contain sodium hydrogen sulfate, an acidic substance that will quickly liberate chlorine from bleach. The acrid fumes of chlorine can then destroy lung tissue, cause the lungs to fill with water, and, in a sense, cause death by drowning. The gas was used for this purpose in World War I. Our mouse-fearing lady almost suffered the same fate as did the French troops at Ypres at the hands of the Germans. Luckily, her neighbor looked in to see how the experiment was going and saved her just as she was about to pass out.

Not everyone who concocts this mixture turns out to be so lucky. Many who have poured bleach into a toilet bowl following an unsuccessful attempt to remove stains with a commercial cleaner have suffered permanent lung damage, and some have even died. No acid must ever be mixed with chlorine bleach — this includes acidic drain cleaners, rust removers, and even vinegar. Drain cleaners can cause all kinds of problems. The most common ones are based on sodium hydroxide, commonly known as lye. They may be sold as solutions or as solid pellets of sodium hydroxide, but products that contain concentrated sulfuric acid are also available. Individually, either vari-

ety of drain cleaner may prove effective, but the two must never be mixed. When these chemicals are combined, they produce a tremendous amount of heat. There have been reports of people who tried to unclog their drains with one type of cleaner followed by a chaser of the other variety. This created heat, which generated steam, which blew the whole corrosive mixture into their faces.

Mixing bleach with ammonia, an ingredient in many window cleaners, can also pose a hazard. Irritating chloramine vapors are released. These are not as dangerous as chlorine, but they are most unpleasant; in fact, the smell people associate with chlorine in swimming pools is not actually chlorine but rather chloramines formed by the reaction of chlorine with urea in the water. Let's not think about why the water contains urea in the first place.

Speaking of swimming pools, disaster may strike if the chemicals required to disinfect pool water are not mixed properly. There are two commonly available chlorinating agents for the treatment of pool water, both of which are usually sold in dry crystalline form. In the water, both release hypochlorous acid, which is the actual disinfecting agent. Calcium hypochlorite is used for short-term protection, and trichloroiso-cyanuronate, also known as stabilized chlorine, releases chlorine over a longer period. These chemicals must be added to the pool water individually. If the dry crystals are mixed in a bucket and water is added, an exothermic reaction, which releases chlorine gas, begins immediately — there is even the possibility of an explosion. The reaction can be so serious that these two substances should not even be stored near each other.

In fact, dry calcium hypochlorite should not be mixed with any combustible substance because it is a strong oxidizing agent. This is just a technical way of saying that it helps substances to burn. Just ask the scoutmaster who decided to teach his charges about proper outhouse techniques. Many outhouses are equipped with a bucket of lime, or calcium oxide. Periodically, a scoop of the stuff is dumped into the hole for smell-control and disinfection purposes. As luck would have it, when the scoutmaster wanted to demonstrate this procedure, he discovered the bucket was empty. He searched around for the main stock of lime and discovered a bag labeled "calcium hypochlorite" in the hut where the swimming-pool supplies were kept. Remembering that lime was calcium something or other, he filled the bucket with the stuff. He then dumped some of the powder down the outhouse hole, and as he walked away the entire campground was shaken by an explosion.

Sewage produces copious amounts of methane gas, which is highly combustible. When the scoutmaster dumped the oxidizing calcium hypochlorite into the hole, the methane exploded. He and his cubs learned a valuable lesson that day about the importance of knowing some basic chemistry. Calcium oxide and calcium hypochlorite are very different substances. When in any doubt, do not mix chemicals. Perhaps Rudyard Kipling put it best about a hundred years ago:

There are those whose study is of smells
And to attentive schools rehearse
How something mixed with something else
Makes something worse.

ZEOLITES TO THE RESCUE

The stench in my car was unbelievable. All I could think about every time I got into the infernal contraption was the classic *Seinfeld* episode in which Jerry's car is saturated with the intense body odor of a parking attendant; no process known to humankind is able to rid his vehicle of the smell. The origin of my problem was different: carrot juice. That's right. I had bought a bottle, placed it on the floor of the car, and somehow forgotten about it. The juice bottle had rolled under the seat, spilled, and begun to ferment. The first sign of a problem was a putrid, yeasty smell. I quickly traced it to the spilled juice and cleaned up the mess, assuming that this would solve the problem. Wishful thinking.

The next day the smell was even worse. In fact, it was indescribable. It seemed as though the fermenting microbes had set up shop in the carpet of the car and were spewing their noxious vapors in industrial amounts: the time had come to put my chemical knowledge to use. First I tried sodium bicarbonate. While baking soda, as it is better known, has a fabled ability to neutralize odors, especially acidic ones, in this case it was like trying to bring down an elephant with a slingshot. With vinegar I fared no better. Next I brought out the activated carbon. This stuff, which is actually charcoal that has been heated to a high temperature in the absence of oxygen, has an amazing ability to draw substances to its surface, and it is commonly incorporated into air and water filters to remove impurities. It didn't make a dent in the stench.

My next idea was to mask the smell, but the fetid vapors just seemed to scorn my attempts to overpower them with dangling car deodorizers and an array of air fresheners. These products simply introduce smells of their own — they can cover the everyday annoying odors, but not the reek from hell.

It was now time to bring out the heavy chemical equipment. Burn the smell! Well, not literally. The more appropriate term is "oxidize" it. Oxidizing agents can rob molecules of the electrons that hold them together, and thus degrade them into simpler substances. Chlorine bleach is a great oxidizing agent. At one time or another, we've all seen how it can shred fabrics if too much is used. By now, I didn't even care about what bleaching would do to the carpet. The smell had to go. I mixed up a strong solution of bleach and sponged it into the battlefield. This should have oxidized the smell and killed the microbes that were responsible for the fermentation, but the only thing I got from this endeavor was the disturbing fragrance of bleach, which lingered for a few days before it was replaced by the original stench. The stinky compounds seemed to be mocking my chemical expertise. Hit me with more, they taunted. I did, with ethylene oxide. This is the stuff hospitals use to sterilize surgical instruments. It has recently become available as an active ingredient in the commercial product known as "Air Sponge." I had achieved success with this product when confronted with several other smell problems, but the ethylene oxide could not penetrate my car stench.

I racked my brain. How on earth could I break down that hellish mixture of organic compounds? Enzymes! I have for a long time been counseling people to use enzyme-based products against odors arising from pet accidents, and they do work. These preparations are available in pet stores, and they can eliminate the smell of urine, vomit, and even cat spray. Enzymes are biologically active proteins that can break down all kinds of organic waste; surely the enzyme army could do the job. But no — this time the army fired blanks.

By now the situation had become desperate. My wife and kids were refusing to get into the car. My pride, my very reputation, seemed to be at stake. I was stymied. My chemical arsenal was depleted. And then I remembered the story about the billy goat: somewhere, I'd read that these animals impart a terrible odor to their handlers, who solve the problem by rubbing their hands with powdered zeolite. Zeolite. Of course! I chastised myself for not having thought of this sooner. I'd certainly known about this material — in fact, a few years earlier I had discussed its smell-eliminating properties on the radio, and I even had a sample of a zeolite product a distributor had sent me at the time. I dug it up and sprinkled it in the car.

But let me leave you in suspense as to the final outcome for a moment while I tell you a little about zeolites. The name derives from the Greek words "zein," which means "to boil," and "lithos," which means "stone": zeolites are "boiling stones." It was way back in 1756 that Swedish mineralogist Baron A.F. Cronstedt noted that certain rocks seem to boil when heated with a flame. These minerals crystallize in the presence of water, which they retain in the pores or channels in their crystal structure. When heated, the water boils out. Dried zeolites are permeated with molecular-sized channels into which water can be reabsorbed. In fact, zeolites are now widely used as antifog agents in double- and triple-glazed windows. And zeolites can trap a variety of other molecules in their porous internal structures as well.

Today, we have a large variety of natural and synthetic zeolites. They are all essentially aluminosilicates, meaning that they are composed of aluminum, silicon, and oxygen. These elements constitute the crystal framework, and their specific relative abundance and bonding patterns determine the size of the channels that run throughout the crystal. Some zeolites, for example, can preferentially trap nitrogen and can therefore be used to separate the oxygen and nitrogen components of air. Others can exchange sodium ions trapped in their structures for calcium and magnesium and thus remove these from water. This effect is known as water softening, and it accounts for the inclusion of zeolites in detergents. Detergents do not work well in hard water — that is, water with a high mineral content. By and large, zeolites have replaced the environmentally un-friendly phosphates in most detergent formulations. The zeolite with the right pore size can even remove undesirable com-pounds such as methyl mercaptan, which can taint instant coffee. This objectionable, skunklike smell is just one of the seven-hundred-odd compounds in the aroma of coffee, but it yields to the right zeolite.

So, if methyl mercaptan could be trapped by a zeolite, why couldn't my horrendous carrot-juice witch's brew? Well, it could: after a couple of powdered-zeolite treatments the stench left my car. Life was worth living again. Stimulated by my suc-cess, I did a little search of the literature to see what other unusual things zeolites can do — and did I ever find one! It seems that while young roosters are eager to mate many times a day, older ones slack off. Putting their minds to this problem, researchers at the Ethyl Corporation discovered that by incor-porating "Zeolite A" into feed they could enhance the roost-ers' desire to mate!

SENSE OR NONSENSE?

GETTING A CHARGE OUT OF
ELECTRICAL NONSENSE

The Vermont farmer couldn't sleep. His heart was pounding so hard he thought it would jump right out of his chest. Unable to bear it any longer, he got out of bed and went outside to walk around. In the dark, he stumbled and fell, landing chest first against the electrified fence he had installed to prevent his cows from wandering off. When he recovered from the shock, the farmer was delighted to discover that his disturbing palpitations had ceased. This inventive man then ran an electrical extension from the fence into his house and treated himself whenever necessary. Much later, he saw a physician for an unrelated problem and refused any advice on his heart condition, claiming he had it "solved."

Upon hearing this story, most people, although perhaps surprised that the farmer did not manage to electrocute himself, would readily accept the validity of his unique "therapy." After all, movies and television shows often portray failing hearts being restored by a shock delivered to the chest. Defibrillation, as this process is known, is actually one of a very few legitimate techniques whereby a patient can benefit from the application of an electric current. But scientifically illegitimate electrical treatments are far more numerous and give rise to some rather shocking anecdotes.

In the late 1800s, miracle after miracle was presented to the American public. Marconi's radio, Bell's telephone, and Edison's lightbulb ushered in the Age of Electricity. If this mysterious force could send sound through the air and dispel darkness, could it not also work its magic on the human body? Scientists began to address this question in earnest, but long before they could cast any light upon it, the quacks and charlatans entered the game. Unencumbered by the need to play by the rules of science, they sparked plenty of interest with their unsubstantiated claims and pseudoscientific lingo.

Galvanic Electric Belts were said to cure "nervous and chronic diseases without medicine." Containing primitive batteries consisting of pieces of copper and zinc separated by blotting paper, they delivered a mild current to the gullible wearer, who would become convinced that the healing process was underway. One of the most popular designs featured loops extending down to the testicles; such a belt was intended to restore "lost manhood," a loss the manufacturer attributed to "the greatest outrage on Nature's sexual ordinances man can possibly perpetrate" — otherwise known as "self-abuse." Perpetrators of this transgression against nature could be identified by the black-and-blue discolorations under their eyes. Fortunately, they could also be reenergized and dissuaded from engaging in such activity by the Galvanic Belt.

And for those who were leery of electrical equipment there was Electric Liniment or pills that "contained 50,000 volts of electricity in a 2 drachm bottle." The only charge patients got out of these nonsensical nostrums appeared on the bill from the quack who sold them the stuff. Surely, the dean of the electrical quacks was Dr. Albert Abrams, a traditionally trained physician who practiced standard medicine, and, for a period in the early 1900s, served as vice president of the California Medical Society. As he approached middle age, Abrams decided that standard medicine did not really suit him, so, in 1909, he invented his own specialty, which he called "spondylotherapy." There was no longer any need to rely on symptoms or stethoscopes in making a diagnosis — Abrams had decided that he could identify the problem by observing how a patient's spine resonated when tapped. After making a diagnosis, Abrams would activate a cure by further drumming on the spine in an appropriate rhythm.

The widespread introduction of electricity was tailor-made for Abrams: now he could put his vibrational ideas on a scientific footing. Diseases, he proclaimed, are caused by a disharmony of electronic oscillations in the body and can be cured by vibrations that have the same frequency as the disease. He invented a device he called "the dynamizer" to diagnose illness by measuring the electronic vibrations in a drop of blood. Diagnosis did not even require the presence of the patient, but it did require a healthy surrogate. Just picture this bizarre scene. A few drops of the patient's blood were treated with a large magnet to "cleanse" them of confusing vibrations and then introduced into the dynamizer. A wire from this contraption was connected to the forehead of a healthy volunteer who stood on a metal plate. Abrams proceeded to tap the surrogate's body systematically until he located an area that somehow resonated with the vibrational frequencies of the blood sample. Thus, the diseased organ was identified (and, incidentally, so was the patient's religion). Abrams then used a second machine, called an "oscilloclast," which he tuned to the vibrational frequency of the disease in order to cure it. Testimonials to the brilliance of Abrams's machines were issued and the money began to flow in.

Noted physicist Robert Millikan called the oscilloclast a device a ten-year-old boy would build to fool an eight year old. The *New York Times* called spondylotherapy a scheme of magnificent absurdity. The American Medical Association produced posters claiming that Abrams's disciples diagnosed nonexistent diseases and then made a fortune by "treating" them. But the good doctor was undeterred by all this, and business flourished. When Prohibition came along, he introduced a gadget that could duplicate the vibrational frequency of alcohol so that Abramsites could get drunk without drinking. More testimonials followed. Finally, public skepticism began to kick in after Abrams diagnosed general cancer and tuberculosis of the urogenital tract in a sample of chicken blood. And interest really waned when Abrams himself contracted pneumonia and died from the disease his oscilloclast was supposed to cure with ease.

Of course, quack electrical devices did not die with him — in fact, they have proliferated lately through the Internet. You can get yourself a Medicomat, which will treat asthma, arthritis, and hepatitis, an Interro device which will diagnose "imbalances" in the body and recommend homeopathic treatments, or a Q-LINK Pendant, which will combat "toxic forms of energy," and which consists of a plastic case, a coil of copper wire, and a computer chip — a bargain at US $129. Then there is the Crystaldyne pain reliever, guaranteed to alleviate pain associated with conditions ranging from arthritis to menstruation. Well, I just had to order one of those. What I got for my fifty dollars of "research funds" was a two-dollar barbecue-grill igniter. All I had to do to eliminate pain was to press the thing against my skin and push the button. The device did come in handy: I used it to replace the nonfunctioning igniter on my barbecue.

ILLUSION, DELUSION, OR SOLUTION?

Living in New York City can undoubtedly be stressful, so it is not surprising that in many pharmacies you'll find a product called New York Stress Tabs. The label describes the tab as a "homeopathic lozenge designed to manage daily stresses related to sleep, work, relationships, travel, hangover, overindulgence, and premenstrual syndrome." The instructions say that a lozenge should be dissolved in the mouth and that the process may be repeated hourly as needed. New York must indeed be a very intense place.

What magical ingredients can these lozenges contain that will so easily take the edge off a crazy day? The label actually reveals the presence of aconite and strychnine — two classic poisons. But not to worry. This is a homeopathic remedy, which means that the ingredients are present in vanishingly small amounts, in fact, they have been diluted to the extent that in most cases there isn't even a single molecule of the original substance left; only some sort of "imprint" or "molecular ghost" remains.

I have a problem with homeopathy. If I am to accept its principles I must cast aside the understanding of chemistry that I have developed over 30 years. Therapy based on nonexistent molecules just does not fit the model. But, of course, I cannot conclude that homeopathy does not work just because its proposed mechanism of action is unacceptable to the current scientific view. After all, it was once widely believed that due to the curvature of the earth, radio transmission across the Atlantic would never be possible because radio waves traveled in straight lines. Then it was accidentally discovered that these waves bounce off the atmosphere. However, before we begin altering our theories about molecules, we have to investigate whether homeopathy really does work. First, a bit of history.

The father of homeopathy, Samuel Hahnemann, was trained in traditional medicine and began practicing in Germany in the late 1700s. He quickly became disillusioned with the treatments he had learned. Bleeding, leeches, suction cups, purges, and arsenic powders seemed to do more harm than good. Hahnemann began to ignore his training and to prescribe a regimen that at the time was quite revolutionary: fresh air, personal hygiene, exercise, and a nourishing diet. Since there was little chance of earning a living by simply recommending this regimen, he began to supplement his income by making use of his fluency in eight languages: he undertook to translate medical texts. While working on one of these translations, he encountered an explanation of why quinine supposedly cured malaria — the substance fortifies the stomach.

Intrigued, Hahnemann took some quinine himself to see if it really had this effect. It did not, but soon Hahnemann began to feel feverish: his pulse quickened, his extremities became cold, his head throbbed. As these symptoms were exactly like those of malaria, he formulated a dramatic conclusion: the reason quinine cured malaria was that fever cures fever. In other words, like cures like. Homeopathy, from the Greek "homoios," meaning "like," and "pathos," meaning suffering, was born.

Hahnemann went further, and began systematically to test the effects of a large variety of natural substances on healthy people. Such "provings" led him to conclude that belladonna, for example, could be used to treat sore throats because it caused throat constriction in healthy subjects. But belladonna is a classic poison. Was homeopathy therefore dangerous? Not at all. Hahnemann had another idea. He theorized that his medications would work by the Law of Infinitesimals: the smaller the dose of a given substance, the more effective that substance would be in stimulating the body's "vital force" to ward off disease.

The dilutions were extreme. "Active preparations" were made by repeated tenfold dilutions of the original extract. Hahnemann was not bothered by the fact that at these dilutions none of the original substance remained; he claimed that the power of the curative solution did not come from the presence of an active ingredient but from the fact that the original substance had, in some way, imprinted itself on the solution. In other words, the water somehow "remembered" the material that had been dissolved in it several dilutions back. This imprinting process had to be carried out very carefully; a simple dilution of the solution was not enough. The vial had to be struck against a special leather pillow a fixed number of times in order to be "dynamized."

Traditional medicine did not take kindly to these peculiar rites. In fact, the American Medical Association was formed in 1846 largely as a reaction to homeopathy; one of its founding goals was to rid the profession of homeopaths. At times, the association's strictures became ridiculous. One Connecticut doctor lost his membership for consulting a homeopath — who happened to be his wife.

Nevertheless, homeopathy did not disappear and is now enjoying a rebirth. People disillusioned with scientific medicine are resorting to homeopathy, gleefully pointing to studies in peer-reviewed scientific journals that appear to show that homeopathy works. But wait a minute. A careful review of these studies yields unimpressive results. In the treatment of some minor conditions, homeopathy does seem to be slightly more effective than a placebo, but this has no practical implication; it merely attracts academic interest. How can there be any positive results at all if there is no active ingredient? Publication bias is a likely explanation. What I mean by this is that if enough studies are carried out, sooner or later some will have to show positive results based on the law of averages. Reporting these while maintaining silence on negative findings can create the illusion of effectiveness.

Recently, the largest-ever review of homeopathic studies was published in the leading British medical journal, the *Lancet*. When all the studies were pooled, homeopathy was shown to have a slight statistical advantage over placebo. But, in the words of the researchers involved: "We found insufficient evidence that homeopathy is clearly efficacious for any single clinical condition." In other words, for all intents and purposes, homeopathy does not work. And this study was carried out by homeopaths. Clearly, no scientific study will derail the advocates of homeopathy. They will keep buying and selling their "cures" for a host of conditions, buttressed by all kinds of anecdotal evidence. Most people just do not realize that the majority of ailments are self-limiting and resolve by themselves.

Homeopaths have even touted a cure for the common cold, something that has stymied scientific researchers. It is based on freeze-dried extract of duck liver, which is diluted to the extent that a single duck can supply the world demand for a whole year. There may not be a goose that can lay a golden egg, but there is, apparently, a duck with a twenty-million-dollar liver.

POP ROCKS AND EXPLODING STOMACHS

Marli Brianna Hughes looks only about three years old, but she is destined to be a TV star. For Marli, you see, has just been introduced by the Quaker company as the new "Mikey" and will be widely featured in the company's television ads for Life cereal. I hope she has a stronger stomach than the original Mikey.

Let's backtrack a little to 1971, when Quaker came out with a commercial that featured a cute little boy who was tasting an

unfamiliar cereal. Downing a couple of spoonfuls, his face reflected pure delight. After awhile, when the ad had run its course, it went off the air, but that's when the stories started. Mikey, the rumors went, had not peacefully retired from his television career: he had died. And what a death. Poor Mikey, it seems, had consumed some Pop Rocks and washed them down with a chaser of soda. His little stomach, unable to handle the pressure from the carbon dioxide released by this volatile combo, had exploded. Could this be?

Pop Rocks were introduced to the American market in 1974. This novel candy was invented by William Mitchell, a research scientist at General Foods, and it was described as "carbonated candy crystals that crackle on the tongue." The product was an instant hit with kids, who got a real kick out of the little pieces of candy that would burst in the mouth. The technology behind Pop Rocks was ingenious. Mitchell had found a way to infuse a melt of sugar, lactose, and corn syrup with carbon dioxide and to retain the gas in the mixture after it had hardened. The sweet ingredients as well as flavors and colors were dissolved in a little water and heated to about 150 degrees Celsius. Application of a vacuum reduced the water content, and carbon dioxide was then pumped in under high pressure.

While the pressure was maintained in the closed vessel, the candy was allowed to cool to a glassy solid. Releasing the pressure then allowed most of the carbon dioxide to escape, a process that cracked the hardened mass into small rocklike pieces. Some of the carbon dioxide, however, was retained in the bubbles that had formed as the candy solidified. Sucking on the little "rocks" resulted in audible pops as this gas was liberated.

But is this truly a killer candy? Of course not: a handful of Pop Rocks contains no more carbon dioxide than a swallow of soda pop. In a worst-case scenario, if coupled with a carbonated soft drink, the candy may produce a burp or two. Yet in spite of its absurdity, the story about the death of Mikey took on a life of its own. General Foods received so many inquiries from worried parents that it had to send out letters to more than fifty thousand school principals explaining that Pop Rocks was an entirely safe product. Even full-page ads in 45 major newspapers and a lecture tour by Mitchell, the inventor, could not alleviate consumer fears. Eventually, the company threw in the towel and stopped producing Pop Rocks. Ignorance had triumphed, and the story of Mikey and his exploding stomach faded away.

Then, a few years ago, a story appeared in several newspapers about another exploding stomach. A silly scare story, I thought — another irresponsible press report designed to grab the reader's attention. Still, I decided to check it out. The story focused on an unfortunate gentleman who had consumed a large Mexican meal before going to bed. He awoke in the middle of the night with his stomach on fire. Surely, he thought, a little sodium bicarbonate would neutralize the excess acid in his stomach and quell those flames. Dissolving a spoonful of baking soda (as bicarbonate is better known) in a glass of water, he quickly drank the mixture down, but instead of experiencing relief he began to suffer severe abdominal pain. The pain became so excruciating that he sought medical help. The diagnosis was a ruptured stomach and emergency surgery ensued.

This struck me as a little far-fetched. Even those students who are just embarking on their chemistry studies know that when baking soda reacts with an acid, carbon dioxide is released — many a science-fair project is based on creating a "volcano" by combining baking soda with an acid such as vinegar. It should therefore come as no surprise that the hydrochloric acid in the stomach can combine with baking soda to release gas, but can it release enough to burst the stomach? Wouldn't the gas just be released as a burp? Hoping to explode another goofy science story, I decided to telephone the victim, whose name had been mentioned in the story. Instead of interviewing some crank, as I had expected, I found myself talking to the editor of *National Geographic* magazine. I learned that his misfortune had, in fact, been accurately reported. To be sure, his was a very rare case. His doctors had afterwards explained to him that because his stomach had been very full when he took the bicarbonate, the expanding gas had pushed the contents around, blocking both the upward and downward openings to the stomach. The building pressure had ruptured the stomach wall; several operations were required to correct the damage.

Although this story is well documented, it is important to realize that stomachs very rarely explode. In the case of the Mexican-food-loving editor, there may well have been some preexisting medical condition that had weakened the stomach. Nevertheless, it seems that it is not a good idea to take baking soda on a full stomach, especially when a number of antacids

— such as aluminum or magnesium hydroxide — that do not produce carbon dioxide are available.

So, if the latter story is true, could the Mikey story be credible after all? Pop Rocks have actually reappeared, this time manufactured by a Spanish company. I bought some. They produced an interesting tingling sensation in my mouth, but that was it — not even a burp. Way too little carbon dioxide to do any damage. Mikey was definitely not done in by Pop Rocks. Anyway, we now we have the ultimate proof. Who introduced to the public Marli Brianna Hughes, chosen from over 35,000 entrants, as the new symbol for Life cereal? None other than John Gilchrist, the original Mikey, very much alive but probably sick of stomaching all the silly rumors about his untimely demise.

STRANGER THAN FICTION

One of the most fascinating aspects of science is that its course is unpredictable. In fact, just about the only thing we can predict with scientific accuracy is that the predictions of psychics will not come true. I would venture to say recent predictions that Madonna will reveal she is Jim Nabors's love child or that Dolly Parton's left breast will explode on a TV special or that President Clinton will admit he's an alien will not pan out. Granted, such predictions are amusing to contemplate, but why look to such silly stuff for amusement when, as the saying goes, truth is even stranger, and often more amusing, than fiction? Yes, there *is* levity in science.

As we're talking about levity, let's start with hydrogen, a lighter-than-air gas. Most people associate it with the Hindenburg explosion or the Challenger tragedy, but would you believe that in Japan they brew a beer replacing some of the

carbon dioxide with hydrogen gas? The manufacturer has offered consumers a lame excuse about reducing the greenhouse
effect by curbing carbon-dioxide emissions, but the real reason
Suiso beer is made this way is that it temporarily gives guzzlers
an unusually high-pitched voice — vocal cords vibrate at a different frequency in an atmosphere of exhaled hydrogen gas.
The Donald Duck-like tones the beer produces go over very
well in karaoke bars, but what goes over even better is the
spectacular fireworks display Suiso drinkers can create by
igniting their own hydrogenated breath. This has led to a rather
dangerous form of entertainment in which participants vie to
see who can breathe the most fire.

One Toshira Otama liked to impress onlookers by downing
15 beers and belching huge amounts of hydrogen. He was
reportedly able to catapult balls of flame across the bar, to the
intense admiration of everyone except the bouncer. This gentleman declared Mr. Otama's dragon act too dangerous after he'd
singed the hair and eyebrows of a patron; the bouncer attempted
to curb the activity. In the scuffle that followed, Otama swallowed his cigarette and ignited the hydrogen gas. He suffered
terrible burns to his esophagus, sinuses, and larynx. Since his
vocal cords were charred, Otama was unavailable for comment, but one suspects he will be looking for less precarious
forms of entertainment in the future.

Otama's exploits may be astounding, but they pale in
comparison to those of Balaram Sharan, a New Delhi yoga
instructor. Sharan stunned a gathering at a press club by sucking 150 milliliters of oil into his bladder through his penis and
then discharging it into an oil lamp. He proceeded to light the
lamp, proving that the emission was indeed still oil. While
Sharan is not a professional performer, his demonstration
undoubtedly has a certain entertainment value. He is actually
in the health field. His theory is that if everyone could perform

this stunt, the world would be free of disease because disease starts in the bladder. I'm not quite sure how one would go about learning this extraordinary skill; furthermore, Sharan has not told us if the oil lavage also works for women, who would appear to have an even greater plumbing challenge to overcome.

The insightful Sharan has, however, made it clear that cleanliness of the bowel is of equal importance. And don't ever let it be said that this yogi is not multitalented. After his oil demonstration, he sucked three liters of water from a bucket into his bowel through a rubber tube and then amazed everyone by spewing the water from his mouth. Presumably, this cleaned out his bowel and made him healthier. I think I would rather eat bran.

Sharan performed his incredible leger-de-rectum totally naked — shame did not disrupt his performance. This makes sense in light of a study carried out recently at the University of Michigan in which volunteers were asked to write a math exam clad only in swimsuits. They worked in private cubicles equipped with a desk and a mirror. Fully clothed women did much better than the swimsuited ones, but men scored higher when scantily clad. It seems that the women were so concerned about how they looked that their mental performance was affected. Why the men did better without their clothes on is not clear; neither is it clear why anyone would undertake such research.

I guess it's just plain old curiosity. The fuel of science. It doesn't always take you somewhere significant, but without it you go nowhere. Scientists will study anything, as long as it is interesting. Like coital frequency and health. A ten-year-long epidemiological study conducted in a Welsh village showed clearly that men who have the most sex live the longest: those who were active at least twice a week had longer and healthier lives than once-a-month performers. Amazingly, there was a dose-response relationship, with good health being directly linked to frequency of intercourse.

These healthy Welshmen, though, may have had an un-healthy effect on their neighbors. Researchers at the University of Cardiff, also in Wales, studied the effect of noisy love-making on neighbors. Basing their conclusions on numerous interviews, they claimed that people were more irritated by the moaning and groaning than, say, a loud stereo. The irritated test subjects said they were stressed because they thought that complaining about coital noise pollution was inappropriate. They were particularly disturbed when the sound effects in-cluded the words "yes, yes, yes!" I kid you not.

There may, however, be a solution to the anxiety created by the overzealousness of others: the rocking chair. Researchers at the University of Rochester demonstrated that rocking reduces anxiety and eases depression. Of 18 elderly patients who rocked for 80 minutes a day, 10 showed striking improvement on tests that measure anxiety and depression. A rocking chair, I suspect, would therefore have come in handy for the New Zealand woman who was talking to a friend when all of a sudden she heard a chicken squawking in the kitchen. This disturbed her, because the only chicken she knew of in the vicinity was the one she had just popped into the oven. Indeed, that is exactly where the sound was coming from. With visions of chicken ghosts dancing in her head, she threw open the oven door and removed the noisy bird. It seems the vocal cords in its neck were still intact, and steam coming from the stuffing caused them to vibrate.

Anxiety-producing stuff, to be sure, but not as unsettling as the experience of a shopper in an Arkansas supermarket park-ing lot. A passerby noticed the lady sitting in her car, clutching her head, not moving. She tapped on the window and asked if anything was wrong. The answer she received was shocking: "I've been shot in the head, and I'm holding my brains in." Paramedics were summoned, and they discovered that the lady was clutching a lump of bread dough to the back of her head. Apparently she had purchased some canned dough, which had exploded in the hot car, making a sound like a shot. The horrified shopper had felt a wad of dough clinging to the back of her head, and became convinced that her brains had been blown out.

Now, tell me, could any psychic have predicted these strange-but-true happenings? Of course not. Psychics didn't even manage to predict the success of Viagra. But let me take a shot at making predictions: I think the radish, chicken-foot, and banana-peel diet will be the new rage; a multilevel marketing company will introduce oxygenated lettuce juice as a cure-all; and a study will show that people who take vitamin supplements eventually die. But I could be wrong. After all, in 1981 someone said that as far as computers go, 640K should be enough for anyone. Who made this pronouncement? None other than Bill Gates.

COLORFUL NONSENSE

When I was young and had a sore throat, my mother would always wrap a scarf around my neck. It felt warm and comforting, but I don't think it had much therapeutic value. That may be because it was the wrong color — at least, that's what the purveyors of the Healing Scarf would undoubtedly suggest. This latest medical miracle caught my eye as I perused the Internet because of its claim to "balance your energies and increase your sense of well-being." Granted, it is a pretty scarf: made of Chinese silk, it features the colors of the rainbow and is "designed to bring all healing colors into your consciousness."

Healing with color? Where did they get this idea? I decided to find out. I'm glad I did, for the winding path this took me down led me to one of the most fascinating characters in the history of scientific quackery. Let me tell you about Dinshah P. Ghadiali and his Spectro-Chrome. Dinshah, as he liked to be called, was born in India in 1873. At least by his own account, he was a remarkable man. He began school at the ripe old age of two and a half; by eight he was in high school, and by eleven he was an assistant to a professor of mathematics at a Bombay college. In his writings, Dinshah claims that he began to study medicine at the age of 14, but then we hear no more about the prodigy's progress in this area — probably because he saw no need to continue this futile undertaking once he had independently discovered the key to health. Color therapy.

Dinshah made this discovery when he cured a young girl dying of colitis by exposing her to light from a lamp fitted with an indigo-colored filter. The therapy also involved giving the patient milk that had been placed in a bottle of the same color and exposed to sunlight. Within three days, the girl was well and Dinshah's career had been launched. He opened Electro-Medical Hall, and there he began to refine his treatment techniques. By the time he traveled to America in 1911, he had a theory — albeit a bizarre one — to go with his colored lights. Every element, he said, exhibits a preponderance of one of the seven prismatic colors. Oxygen, hydrogen, nitrogen, and carbon, the elements that make up 97 percent of the body, are associated with blue, red, green, and yellow. In a healthy person these colors are balanced, but they fall out of balance when disease strikes. The therapy is simple: to cure a disease, administer the colors that are lacking or reduce the colors that have become too brilliant.

To enact this therapy, Dinshah developed the Spectro-

Chrome, a box with a lightbulb in it and an opening that could be fitted with various colored filters. This he sold accompanied by the Spectro-Chrome Therapeutic System guide, which detailed the appropriate colors to shine on a given patient. Green light, for example, was a pituitary stimulant and germicide, while scarlet was a genital stimulant. Any disease or condition, Dinshah insisted — save broken bones — was amenable to color therapy. He also maintained that the Spectro-Chrome was especially suited for use by intelligent people, because "drugs quickly upset the nervo-vital balance of persons of high mental and spiritual development." A pretty clever ploy — the gullible, thinking themselves to be intelligent, ate it up.

At the time, many people found his arguments about the benefits of color therapy to be quite convincing. After all, they knew that premature babies were treated with blue light to cure them of jaundice, that sunlight was needed for the synthesis of vitamin D in the body, and that plants absolutely required light for growth. Add to this the fact that chemists had shown that elements, when heated, emitted different colors of light, and Dinshah's preposterous notions started to make sense. His slogan, "No Diagnosis, No Drugs, No Surgery," also sat well with a public largely unsatisfied with available medical care. The idea of a noninvasive therapy and the promise of a cure for virtually any ailment were very appealing.

Of course, it wasn't long before Dinshah ran into trouble with the establishment. He was labeled a fraud and a charlatan by the American Medical Association but managed to portray himself cunningly as a humanitarian who was being persecuted by moneygrubbing, ineffectual, jealous physicians. To protect himself legally, Dinshah came up with some incredible lingo. He didn't talk of "cures," he spoke of "normalating" the body. Instead of "treating" patients, he would "restore their Radio-Active and Radio-Emanative Equilibrium." This he would do with his light exposures, or "tonations." Tonations would be carried out with the patient lying with his head to the north, so as to align the earth's and the body's magnetic fields. Dinshah also designed Spirometer Rods to measure the pressure difference between the two nostrils and thus to determine at what time of day tonations should be carried out with the aim of taking full advantage of the body's natural tides. Special thermometers applied to the bare skin above the organs would determine whether a condition was acute or chronic and what kind of light therapy was needed. It would be hard to imagine a more convoluted and irrational form of therapy.

In 1931, Dinshah had run-in with the law over the Spectro-Chrome. It wasn't his first legal skirmish — six years earlier, he had been arrested for transporting a 19-year-old girl, his secretary, across state lines for immoral purposes. (Perhaps he had been overexposing himself to scarlet light.) He spent four years in jail. But now he was arraigned on second-degree grand-larceny charges after a former student complained to officials that the Spectro-Chrome did not perform as promised. In defending himself, Dinshah trotted out numerous satisfied patients, including — incredibly — several physicians. In fact, a surgeon, Kate Baldwin, claimed that she had successfully treated glaucoma, tuberculosis, cancer, syphilis, and a very serious burn case with Dinshah's device. The government had experts testifying that the Spectro-Chrome was merely an ordinary lamp and that any successes were due to the placebo effect. Ultimately, the prosecution could not prove the intent to defraud, and Dinshah was found not guilty. He went back to selling more Spectro-Chromes, now claiming that he had been vindicated by a court of law.

After the passage of the Food and Drug Act of 1938, which gave the FDA some teeth in regulating therapeutic devices, the government again began to assemble evidence against Dinshah.

Finally, in 1945, he was charged with introducing a misbranded article into interstate commerce, a violation of the criminal code. Once again, he trotted out his satisfied patients, but this time there were no supporting physicians. His fate was virtually sealed when a star witness, whom Dinshah had "cured" of seizures, had a fit on the witness stand. The prosecution called a witness whom Dinshah had repeatedly profiled in his advertising as having been cured of paralysis; she could not take a single step when the master urged her. Another witness described how he had contacted Dinshah after his diabetic father had lapsed into a coma and was simply told to shine a yellow light on him. He did, until his father died. And, finally, the court heard how the celebrated burn victim described as the recipient of a miracle cure by Dr. Baldwin in the previous trial had, in fact, succumbed to her injuries.

Dinshah was heavily fined, his books and lamps were seized, and he was put on five-year probation. The day after his probation ended, he was at it again. This time, he founded the Visible Spectrum Research Institute and sold lamps labeled as having "no curative or therapeutic value." He strenuously implied in his literature that this was only a means of keeping the FDA dogs away — just meaningless legalese that David had to resort to in his eternal battle with Goliath. In 1958, the government obtained a permanent injunction against shipping Spectro-Chromes across state lines, but the persistent Dinshah kept selling the things in New Jersey. After his death in 1966, his sons took over and managed to have the Dinshah Health Society of Malaga registered as a nonprofit, scientific, educational, tax-exempt organization. The society still sells all kinds of light-therapy books, including a history of the Spectro-Chrome by Dinshah himself in seven cloth-bound volumes priced at $220. You can also buy instructions for building an inexpensive Spectro-Chrome from a lightbulb, cardboard, and colored plastic sheets. Apparently, the society does not sell the finished product, but another company on the Internet does advertise Color Light Therapy Lamps "as recommended by Dinshah." These look suspiciously like theater spotlights with colored gels.

It seems that today ignorance about the nature of light, about disease processes, and about how the body functions still abounds; the gullible continue to be victimized. The colorful Dinshah may have lived in what we look upon as enlightened times, but his pseudoscientific ideas smacked of the Dark Ages. I don't know whether he would have approved of the Healing Scarf, but I suspect it would have been right up his alley. The scarf though, I must admit, is so nice that I bought one. It works — it keeps my neck warm.

"Where's the Aura?" Asks Emily Rosa

Rarely does a report published in the *Journal of the American Medical Association* generate as much publicity as did one entitled "A Close Look at Therapeutic Touch." Then again, rarely is the author of a paper appearing in one of the world's leading scientific publications an 11-year-old girl; and rarely is a grade-four science-fair project the subject of a research article. How did this article get into the prestigious journal? The same way any other scientific report does: it held up to the scrutiny of expert reviewers.

The author of the piece was Emily Rosa. With help from her mother, a nurse, she wrote a description of how she had designed and conducted a simple experiment to test the claim of therapeutic touch (TT) practitioners that they are able to detect the "energy aura" that surrounds a human body. These practitioners insist that they can feel the energy field when they move their hands above a patient's body and that they can even reconfigure it to correct "imbalances." Conditions they claim to be able to ameliorate range from arthritis to Alzheimer's disease.

Emily tested 21 practitioners in a straightforward manner. She concealed herself behind a screen and asked the practitioner she was testing to put both hands through the screen. She then hovered her own hand above one of her subject's hands at a distance the TT practitioners had all previously agreed was ideal for sensing the energy field. All the subject had to do was report which hand was sensing Emily's presence. The results were no better than those that would have been obtained from random choice alone.

Not surprisingly, Emily's report sent the proponents of therapeutic touch into mental gyrations. Dolores Krieger, a professor of nursing at New York University who had pioneered TT in 1973, came out with guns blazing: the study was invalid, she said, because one doesn't just go into a room and perform TT, and, furthermore, the healer's hands have to be moving all the time to detect the "aura." But what Krieger didn't acknowledge is that the TT practitioners Emily tested had agreed before the experiment that the conditions were acceptable and that they would be able to feel the field generated by Emily's hand.

Professor Krieger's objections are understandable, given that she has parlayed therapeutic touch into a huge business. There are close to fifty thousand TT practitioners in North America, and TT is taught in many universities. Hospitals often hire these therapists at rates of seventy dollars an hour to balance energy fields around patients. Krieger claims this is well justified and points to the "hundreds" of supporting studies in the scientific literature. When I actually examined these studies, however, I could see that they fall far short of proof. Most of them can be far more readily criticized than Emily Rosa's science-fair project.

While Emily's study showed that TT does not work the way its proponents claim, it certainly did not invalidate the practice. There is no doubt that many people find relief when they believe that their energy fields are being manipulated. When a patient feels better, the why becomes of secondary importance. If therapeutic touch works through the placebo effect, then so be it.

Dolores Krieger, of course, believes that there is more to TT than the power of suggestion. She first became interested in the subject in 1971 when she was a member of a research team created to study the healing abilities of the remarkable Oskar Estabany, a former Hungarian cavalry colonel. Estabany claimed that he could heal horses and people just by the laying on of hands. There were enough testimonials by the 1960s to prompt a study, conducted by Bernard Grad at McGill University. This study appeared to establish that the healer could control the rate at which goiters grew in rats deprived of iodine by placing his hands around their cage for 15 minutes twice a day. In the 1971 study Krieger took part in, Estabany appeared to be able to affect patients' hemoglobin levels. All of this was enough to sell her on the idea of healing by hand motion, and she soon discovered that she could unblock patients' congested energy (even though this phenomenon is unmeasurable by any known device) and infuse them with her own. To her credit, Krieger only works with medical doctors and claims only to be able to effect relaxation and pain relief: she does not talk of miracle cures.

Krieger was obviously not the first to suggest that the human body comes equipped with some sort of spiritual energy that governs health. Indeed, this is one of the most ancient concepts in what we today call alternative medicine. The Chinese have long believed in the mysterious life force called "qi" (pronounced "chi"), which travels through the body's "meridians" and which, when it becomes imbalanced, triggers disease. Correction of this imbalance through acupuncture, patterned breathing, or diet affords relief. The age-old practice of Ayurvedic medicine in India is built upon similar ideas. The human body is seen to be made up of energy elements called "doshas," which operate through body channels; the proper flow of these elements is critical to good health. Neither of these belief systems has any basis in anatomy, and they arose because in China and India dissections were forbidden, leaving the physical workings of the body shrouded in mystery. Healing, therefore, had to be based on metaphysical beliefs.

These beliefs have often proven to be quite potent, and they have withstood the test of time. Their success is probably due to a combination of the power of the imagination and the fact that many illnesses are self-limiting or psychosomatic; any other explanation would force scientists to swallow pretty hard. How, if not for the power of suggestion, could we explain the healing abilities of George Chapman, an uneducated Englishman with no medical knowledge? Chapman claimed to have been contacted by the spirit of a diseased physician who taught him to go into a trance and operate on the spirit body of a patient using invisible surgical instruments. Not only did Chapman's patients eagerly give testimonials to his proficiency, but they also maintained that they had felt the twinge of the scalpel and the drawing together of the spiritual flesh after the "operation."

Wilhelm Reich, a psychoanalyst who trained under Freud, did not look to spirits to learn about the body's health-governing energies: he looked to outer space. Reich believed that he was the product of a relationship between an alien and an earth woman, and that his unusual background permitted him to see not only the body but also the universe as governed by "orgone energy." He derived the term from "orgasm," which he explained is the ultimate expression of this form of energy. Illness, Reich believed, is due to a deficiency of orgone, which he could remedy by placing the patient in an "orgone accumulator," a box about the size of a phone booth with no mechanical or electrical components. The testimonials poured in.

But not all orgone is good: Reich warned that some UFOS are propelled by orgone motors (one wonders what the aliens are doing inside those UFOS — and for how long — in order to generate the necessary orgone), and that Deadly Orgone Energy accumulates in the atmosphere causing disease on earth. Reich obligingly made a device called a Deadly Orgone Buster available to help rid us of this scourge.

Believe it or not, orgone promoters are still with us. One of them sells an orgone generator for the home that is even capable of remote energizing as long as the patient carries a "transfer disk" in a pocket. This Internet advertiser offers proof of the device's efficacy: just download a "transfer diagram" and hold your hand three inches above it to experience a sensation of warmth or a cool, gentle breeze. I tried it. I felt no orgone. The only thing I felt was silly.

THE BOTTOM LINE

GREAT MOMENTS IN URINE HISTORY

These days, urine doesn't often make news. Every once in awhile we'll hear how urine analysis was used to prove an athlete was using steroids or how a drunk in New York electrocuted himself by urinating on the subway tracks. We may give some thought to urine if the cat sprays the carpet or a female dog's urine attracts all the male dogs in the neighborhood. In days gone by, however, there have been several occasions when urine could well have made headlines.

One day in 1669, the *Hamburg Daily* might have led off with the banner "Local Alchemist Invents Cold Fire." Hennig Brandt, like other alchemists of his time, was driven by the desire to make gold and to discover the secret of life. The two were not unrelated. Gold was considered the eternal metal; it does not corrode or tarnish. If the secret of its immortality could be discovered, perhaps it could be applied to humans and make them immortal as well. Brandt searched for this dual secret in urine. Perhaps its yellow color is actually due to the presence of gold, he theorized, and he began to look for methods of extracting the precious substance. He also knew that urine derives from blood, and that blood is essential to life. It therefore seemed reasonable to think that some of the life-giving properties of blood may be found in urine. Brandt collected a large quantity of urine and attempted to concentrate it by boiling it down and then distilling the vapors.

He watched in great anticipation as the vapors condensed. At first, he must have been sorely disappointed, because no gold formed. This letdown was undoubtedly followed by elation as the waxy white substance that now coated his flask began to glow eerily in the dark — it certainly wasn't gold, but could it be the much-sought-after "elixir of life"? Since Hennig is no longer with us, it obviously wasn't, but the strangely glowing substance did confer a certain immortality on the inquisitive alchemist, for Hennig Brandt will forever be known as the discoverer of the element phosphorous.

It didn't take Brandt long to realize that this new substance did more than just glow in the dark. When the paste dried out, it burst into flame. Furthermore, the substance could be stored safely under water and used to produce fire whenever necessary. This was indeed a momentous discovery, because at the time the only way to start a fire was with a flintstone. Brandt tried to keep his method of preparing the exciting substance a secret, and he even managed to capitalize on his discovery by selling the method to a few people. By 1737, however, the French government had studied the process and published a report on it. Brandt's secret was out.

The chemistry turned out to be relatively simple. Urine is a solution of body wastes, and among these are found a variety of phosphates, inorganic compounds in which phosphorous is bound to oxygen. When phosphates are heated in the presence of carbon, this element strips the phosphates of oxygen, forming carbon monoxide and leaving behind elemental phosphorous. Brandt formed the required carbon by heating urine to a high temperature, and this process converted the organic components of the urine to charcoal, which is essentially carbon. This is not unlike the conversion of wood, through burning, to charcoal.

By the end of the eighteenth century, phosphorous had been put to use. The first match had even been developed. It was a rather simple device, consisting of a piece of paper tipped with phosphorous and sealed in a glass tube. When the tube was broken, the phosphorous would come into contact with the air and ignite. Soon there were improvements. It became possible to purchase a sulfur-tipped splint paired with a small bottle of phosphorous; when the splint was dipped into the bottle, the phosphorous would ignite, lighting the sulfur, which would in turn light the paper.

It then didn't take long for the "strike anywhere" match to be developed. This was made by taking a wood splinter; coating the tip with phosphorous, sulfur, and potassium chlorate; and dipping it in glue. The glue prevented air contact until it was rubbed off on an abrasive surface, at which point the phosphorous would ignite. This reaction would be enhanced by the release of more oxygen from the potassium chlorate on the match until a temperature was reached sufficient to ignite the sulfur. This, in turn, would cause the wooden splint to burst into flame.

If a wooden splint could be made to burst into flame with phosphorus, why not a person? After all, man's ingenuity knows no bounds, especially when it comes to creating novel methods of warfare. Soon phosphorous-containing shells and bombs appeared. These would rain fire down on the enemy, the tiny pieces of phosphorous igniting clothing and burning flesh. When phosphorous burns, it combines with oxygen to form phosphorous oxide, which appears in the air as a dense white smoke. This property of phosphorous was exploited in the creation of "smokescreen" bombs.

But where was all the phosphorous required for these uses to come from? Certainly not from boiling urine on an industrial scale. By the early 1800s, chemists had determined that bones also contained phosphates. Animal bones were used first, but when it became apparent that there were not enough of these, battlefields were scoured for human bones. It was only upon discovery of massive deposits of phosphate rock at various places around the globe that phosphorous finally became readily and cheaply available. Today, phosphorous is manufactured on a massive scale by heating phosphate rock with carbon in a process reminiscent of Hennig Brandt's early efforts.

There were several problems with the "white phosphorous" produced in earlier times. First, it was poisonous, and many people working with the substance died from "phossy jaw." This condition was caused by phosphorous vapors entering the body through decayed teeth and destroying bones — the jawbones were affected first. The other problem was that phosphorous ignites so easily. Both of these difficulties were overcome when it was realized that white phosphorous could be converted into a much safer form, known as "red phosphorous," if it was heated in an atmosphere of nitrogen or argon. The first safety matches were made by gluing some sulfur and potassium chlorate to the head of a wooden splint. Provided on the box they came in was a striking surface coated with ground glass to provide friction and with red phosphorous for ignition. The phosphorous would light the sulfur, which, in the presence of the oxygen released by the chlorate, would burn and ignite the wood.

The last improvement in match manufacture evolved from the idea that sulfur and phosphorous can be made to react and thereby produce a compound known as phosphorous sesquisulfide. This substance is not toxic and it is unaffected by the air: it does, however, ignite when its temperature is raised through friction. Modern strike-anywhere matches therefore have a tip of phosphorous sesquisulfide as igniting agent, potassium chlorate as oxidizer, and powdered glass as heat generator — all held together with glue. Modern safety matches have a head covered with potassium chlorate, which is struck against a surface covered with red phosphorous and antimony sulfide. The chlorate provides the oxygen needed to ignite the phosphorous, which ignites the sulfide, which ignites the match. Incidentally, caps for toy guns make use of the same combination of elements — the striking of a mixture of sulfur, red phosphorous, and potassium chlorate causes a mini explosion. Brandt surely never dreamed that his urine would achieve such fame. But our look at the contributions of urine to science is not yet over.

Chemistry can be divided into two basic fields of endeavor: analysis and synthesis. Analytical chemists determine which fundamental substances a given material is composed of, while synthetic chemists assemble compounds from molecular building blocks. The alchemists made attempts both at analysis and synthesis, but we can hardly refer to their secretive, bungling, and often chaotic experiments as science. By the middle of the eighteenth century, a new breed of scientists was emerging. True scientists such as Boyle, Priestley, Lavoisier, Newton, and Dalton became interested in unraveling the secrets of nature, and their sole motivation was curiosity. They wanted to know what their world was made of and how things worked.

The prevailing opinion at the time was that all matter could be divided into two categories — organic and inorganic. Rocks and minerals, which were obviously lifeless, were said to be inorganic, whereas those substances that could be isolated from living systems were referred to as organic. Furthermore, these organic materials, such as opium from the poppy or quinine from the cinchona tree, were thought to possess a vital force that could never be duplicated by humans.

While there was no doubt that chemists could manipulate inorganic substances — for centuries they had been extracting metals from their ores and turning them into swords and ploughshares — making organic substances from inorganic ones was deemed impossible. If one wanted vanilla flavoring, one had better know where to find a vanilla bean.

All of this changed suddenly and dramatically in 1828. Once again, the focal point was urine. As early as 1773, French chemist Hilaire-Marin Rouelle had isolated a white crystalline substance from urine, which appropriately came to be called urea. Since this material had been synthesized by the human body, it was obviously "organic" and could not be made in the laboratory — or at least it was so believed until that monumental day in 1828 when German chemist Friedrich Wohler carried out some tests on a white crystalline material he had made by heating a mineral substance called ammonium isocyanate.

When he treated the crystals with nitric acid they "produced at once a precipitate of glistening scales." It seemed to Wohler that he had seen this kind of an effect before — he racked his brains until he remembered that the natural urea he had once worked with showed similar behavior with the acid. He immediately procured some urea and demonstrated that it was identical to the white crystals made by heating his "inorganic" ammonium isocyanate. Wohler wrote to his former professor Jacob Berzelius in great excitement: "I must tell you that I can make urea without the use of kidneys or of any animal, be it man or dog." Indeed, in one fell swoop he had made urea, destroyed the theory that organic substances possessed some vital force, and built a permanent bridge linking inorganic and organic chemistry. All of a sudden, the potential of chemistry seemed limitless: if urea could be made in the laboratory, so, perhaps, could quinine, or vanillin, or a whole host of substances that didn't even exist yet.

For chemists, Wohler's synthesis of urea buried the idea of a difference between synthetic and natural substances. Since synthetic urea was in every way identical to the natural substance, it became clear that the properties of a specific compound depended only on its composition and not on its ancestry. Wohler's urea was every bit as good a fertilizer as the natural variety because it was identical to it. It is interesting that more than 150 years after Wohler's feat, the notion that natural is somehow superior to synthetic is still subscribed to by a significant portion of the general public. Many still believe that natural vitamin C, for example, contains some intangible property that makes it better than synthetic ascorbic acid.

Wohler's accidental synthesis of urea was soon followed by another important chemical development involving urine. German chemist Karl Ludwig Reichenbach was interested in isolating chemicals from beechwood tar. There was plenty of this stuff around, because beechwood was used to make the charcoal required for the factories that smelted metals from their ores. Reichenbach, in fact, became the first chemist to prepare creosote, the liquid obtained from wood tar by distillation. The rather disagreeable aroma of creosote gave Reichenbach an idea. His house was surrounded by a wooden fence, which — much to his annoyance — often received the attention of the male dogs in the neighborhood. In an effort to discourage leg raising around his property, the inventive chemist painted the fence with creosote, but the dogs did not share Reichenbach's opinion about the foulness of the smell and kept adorning the fence with their urine.

The failure of creosote to act as a dog repellent, however, resulted in an important scientific discovery. Reichenbach observed that a blue color became visible where creosote, wood, and urine came into contact. He soon isolated a blue dye and named it "pittacal," from the Greek words for "tar" and "beautiful." Reichenbach went on to prepare pittacal in a pure state, and he even tried to sell it as a commercial dye. He met with little success in this area, but pittacal maintains an important place in chemical history as the first synthetic dye-stuff. This was some 25 years before William Henry Perkin's celebrated discovery of mauve.

We have actually known that urine can play an important role in color formation since time immemorial. Primitive dyers understood that certain substances help natural dyes adhere to fabrics. These materials came to be known as "mordants," from the Latin word for "bite." For example, when cotton or wool is simmered in water with onion skin nothing much happens, but when a little alum is added to the solution, the fabric is dyed. The alum helps the natural pigments in the onion skin "bite" into the fabric.

The discovery of the mordant effect was undoubtedly an accidental one, probably occurring when people observed that the addition of club mosses to dyeing solutions improved the colors that could be achieved. These mosses are now known to be good natural sources of alum. Such observations surely triggered further experimentation with readily available substances — and what is more readily available than urine or dung? Both of these worked. Roman ladies actually used mullein leaves steeped in urine to make a yellow hair dye. In India, cotton is still mordanted with cow dung. Before the discovery of synthetic indigo, one method of dyeing with natural indigo extract involved steeping the fabric in heated urine. During the nineteenth century, as chemical knowledge and laboratory techniques became more refined, attention naturally turned to the mechanism through which urine achieves its color-enhancing effect. Textile manufacturers obviously had a vested interest in this project.

Adolf Schlieper was a German textile manufacturer who had gained some experience in chemistry by working in the laboratory of Justus von Liebig, one of the leading chemists of the era. Schlieper had worked on uric acid, a white crystalline substance that Scheele had isolated from urinary stones almost a hundred years before. Uric acid is found in small amounts in the urine of all carnivores and is the major component in the excrement of birds, scaly reptiles, caterpillars, and — by some quirk of evolution — Dalmatian dogs. Deposits of uric acid in the joints cause the painful condition known as gout.

Schlieper's work on uric acid never amounted to very much. His main contribution to science was that he gave some of the chemicals he had been working with to another young German chemist, Adolf von Baeyer. His interest aroused, Baeyer began to carry out research on uric acid and its derivatives; it was one of these derivatives that brought Baeyer lasting fame. Starting with uric acid, he prepared a brand-new white crystalline compound, christening it "barbituric acid." There are several theories as to why Baeyer chose this particular name. Some contend that the compound was named after a Munich waitress who had often provided the raw research material. Others say that the discovery took place on St. Barbara's Day. Baeyer himself implied in his lectures that at the time he was in love with a "Miss Barbara" (this, of course, does not negate the waitress story).

Although the origin of the name may be contentious, there is no doubt that Baeyer's discovery of barbituric acid lay the foundations for the development of one of the most important classes of medicinal drugs: the barbiturates. The importance of these compounds is based on the fact that they are central-nervous-system depressants and can induce effects ranging from mild sedation to deep sleep. Barbiturates are used in many prescription sleep medications and are also widely employed in surgical anesthesia.

Baeyer never noted any sedative effect of barbituric acid, but this is not surprising because barbituric acid itself has absolutely no hypnotic properties. It was Emil Fischer, Baeyer's most famous student, who finally discovered that a derivative of barbituric acid, known as barbital, induces sleep. Almost fifty years after Baeyer's first synthesis of barbituric acid, Fischer, working with physician Joseph von Mering, demonstrated that a dog receiving a single injection of barbital would fall asleep. In light of this effect, Fischer renamed the substance "Veronal," after Verona, Italy, which he considered to be the most restful city in the world. Today, when anticipating surgery, we can rest easy because Adolf von Baeyer's interest in urine eventually led to the right chemistry.

Fart Proudly

Benjamin Franklin was a practical man. He gave us the Franklin stove, the lightning rod, and bifocal glasses: science was to be used for the benefit of mankind, the great inventor maintained. Academic societies that emphasized obscure theoretical discussions often fell victim to Franklin's witty wrath. In 1783, the Royal Academy of Brussels posed a question to its members that had to do with the number of times a certain geometric shape could be inscribed inside another, larger shape. This kind of philosophical conjecture did not sit well with Franklin, and when the academy implied that the solution to the problem could actually have practical applications, Franklin put his satirical pen to paper.

He was happy, he wrote to the academy, to hear that the

institution was finally dealing with matters that had the potential to improve society. And, this being the case, he had a suggestion for a future project of great practical significance that he hoped the academicians would consider. Could the academy tell us how we can eliminate the disturbing fragrance of human gaseous emissions? Franklin noted that all humans produce gas, but most people take great pains to hold back their eruptions lest they be accused of unsociable behavior. This type of restraint can cause bloating, colic, and constipation, he continued. Therefore, he was challenging the learned academy members "To discover some drug wholesome and not disagreeable, to be mixed with our common Food, or Sauces, that shall render the Natural Discharges of Wind from our Bodies, not only inoffensive, but agreeable as Perfumes."

Franklin even had some suggestions as to how the project might be approached. Since it was common practice to throw lime into toilets in order to dispel nasty odors, perhaps the academy's scientists could come up with some sort of lime solution that humans could ingest. Or, failing this, what about the possibility of adding pleasant fragrances to food to improve the scent of the emissions? Indeed, in the fashion of wearing perfume to please others, would it not be possible to aromatize flatus to make it not only socially acceptable but also desirable — the scent of lily for some occasions, musk for others?

Franklin obviously had his tongue firmly planted in his cheek — he certainly did not regard flatulence as a major human affliction. The Royal Academy of Brussels never responded to the Franklin challenge, but modern science has. It has broken the silence about this once-taboo topic because after all, everybody does it. Fart, that is. Oh yes, the term is quite acceptable (it derives from an ancient Greek word for "breaking wind"; even Chaucer used it in *The Miller's Tale*), but anyone bothered by it should just consider it an acronym for "flatus advanced by rectal transport."

He who breaths and eats, farts. So does she. To put it simply, gas production is a direct consequence of swallowing air and food. The mechanism of human ventilation is actually quite well understood. The earliest work in this area was carried out by a Frenchman named Magendie, who, in 1816, examined the intestinal gas of newly executed criminals. His identification of carbon dioxide, nitrogen, and methane became the cornerstone of the science of flatology. Magendie, however, missed the most interesting intestinal gas, hydrogen, and was also unable to determine the origin of the odiferous effluents.

Chemistry has now come to the rescue. These days, an instrumental technique called gas chromatography makes the analysis of unknown gases quite easy. Though the raw research material is quite abundant, the actual collection of same requires a little imagination and daring. Dr. Albert Tangerman, a Dutch gastroenterologist, provided six subjects with gas-collection syringes, which they were to press against their rear ends when an outburst was imminent. The gases responsible for the ill wind turned out to be present in very small concentrations — actually less than 1 percent of the total volume —

but there were some 250 ghastly components, with hydrogen sulfide, methanethiol, skatole, dimethyl sulfide, and dimethyl-disulfide responsible for most of the odor.

Where do these gases come from? Some come from swallowed air, but the bulk — that is, the hydrogen, methane, and carbon dioxide — are produced by the action of bacteria on sugars, starches, and fiber in the colon. Whatever components of these dietary staples are not digested and absorbed into the bloodstream as they pass through the small intestine serve as dinner for gut bacteria. The products of bacterial digestion are the aforementioned gases, but individuals have unique bacterial flora in the colon and therefore produce fart patterns as unique as fingerprints. Methane, for example, is detected only in about a third of humans. Caucasians and Blacks produce more than Orientals. It is also noteworthy that methane production is lower in those with Crohn's disease and higher in those with colon cancer. The potency and frequency of emissions is governed mostly by the diet. Certain carbohydrates, such as raffinose, stachyose, and verbascose, present a significant problem because humans lack the enzyme necessary to digest them. Hence the potency of beans, cabbage, cauliflower, and broccoli. Lactose, better known as milk sugar, as well as pectins and beta-glucans in oat bran, are notorious gas producers.

We actually know a surprising amount about gas production and its effects because various practical concerns have stimulated research in this area. For example, during World War II, the development of fighter planes capable of reaching very high altitudes presented a novel problem. As the planes climbed, pilots began to experience abdominal cramps. Every student of elementary physics knows that gases expand as pressure is reduced, and in this case it was the expanding gases in the pilots' guts that caused bloating and pain. Responding to the problem, the US Air Force explored the nature of the foods that caused the greatest discomfort, and concluded that beans and legumes were incompatible with flying. Allied pilots' avoidance of these foods may even have been partially responsible for their success against the Luftwaffe: the Germans were fond of sauerkraut, which is an extremely potent gas producer.

One of the greatest boosts for flatus research was the advent of the Space Age. The control of the internal atmosphere in a space vehicle was a formidable challenge. Removal of the carbon dioxide exhaled by the astronauts was naturally a concern, but some scientists also focused in on how to deal with the buildup of methane and hydrogen in the capsule. These human emissions are highly flammable and were deemed a potential hazard in the oxygen-rich atmosphere of the spacecraft. Accordingly, the astronauts' diet was specially formulated to exclude gas-producing ingredients.

Such attention to detail has prevented embarrassing accidents in outer space, but inner-space problems persist. Intracolonic explosions have actually occurred, sometimes with dire consequences. Growths in the colon, called polyps, are routinely removed to reduce the risk of colon cancer. The procedure is a relatively simple one and involves inserting a snare-equipped probe into the colon. The minor bleeding that usually accompanies the removal of the polyp is stopped by electrocauterization — and therein lies the problem. Prior to the procedure, the patient's colon has to be emptied, and this is usually accomplished by having the patient drink large volumes of a laxative solution. Laxatives work precisely because they are not absorbed by the body and are thus quickly propelled through the digestive tract. A solution of mannitol, for example, will provide a very clean working environment for the physician, but mannitol also serves as food for intestinal bacteria and therefore promotes the formation of hydrogen gas. If any of this gas is present when electrocauterization is performed, both patient and doctor could be in for a nasty shock. Indeed, there have been cases where intestines have been ripped apart and doctors blown clear across the room. When there is a suspicion of hydrogen gas buildup in the colon, flushing with carbon dioxide before the procedure can prevent a great deal of misery.

For doctors, intestinal gas can be an occupational hazard; for lawyers, it can be a financial windfall. Witness the story of the cashier in a Portland, Oregon, grocery store who was accused of tormenting another employee by willfully and repeatedly passing gas. The victim sued for "severe mental stress and humiliation." During the trial, the plaintiff described how the perpetrator of this heinous crime would "hold it in and walk funny to get to me." The clever defense attorney, however, used the argument that breaking wind is a form of free speech guaranteed by the First Amendment. The judge agreed, and the prosecutor's case fizzled out.

Le Petomane

The celebrated performer, dressed magnificently in waistcoat, red breeches, white stockings, and black patent-leather shoes, strode proudly to the center of the Moulin Rouge stage. The capacity crowd was thrilled to see, at last, the most famous French entertainer of the Gay Nineties. Not even the renowned Sarah Bernhardt had as great a public appeal or commanded as high a fee as did Joseph Pujol.

Pujol was a musician of sorts, but he played no instrument. Rather, he himself was a musical instrument — a wind instrument. This illustrious entertainer had the ability to suck air into his body by relaxing his abdominal muscles and then expel the air at will by controlling his rectal sphincter. The unique elasticity of this particular part of Pujol's anatomy allowed him to produce sounds ranging from a clap of thunder to the ripping of cloth. It has been said that he elevated the passing of wind to an art form.

Spectators would howl gleefully when "Le Petomane," as Pujol came to be known, performed a series of wind-passing sound effects: his interpretations of the sonic booms produced by bricklayers, the apologetic tones of nuns, and the barely audible staccato bursts released by brides on their wedding night usually brought down the house. The act ended with Le Petomane blowing out a candle in his unique fashion.

Joseph Pujol was indeed a scientific curiosity. While still a young boy, he discovered his "talent" one day at the beach. He held his breath, put his head under the water, and was shocked to feel a cold, penetrating sensation in his abdomen. Rushing out of the water, young Joseph was astonished to find water rushing out of him; he soon learned that his body could be made to act as a giant pipette, sucking in and releasing water at will. Then came the formidable discovery that he could also inhale and expel air in this extraordinary fashion, and so was born what was arguably the most amazing novelty act of all time.

Pujol sold the act to the manager of the Moulin Rouge in his inimitable way. Having brought a basin filled with water into the gentleman's office, he proceeded to empty and then refill the vessel while sitting upon it. The bewildered manager was also treated to a selection of sound effects and to a rendition of "Au clair de la lune" played on a flute in a decidedly original fashion. Needless to say, Pujol got the job.

Le Petomane became the toast of Paris. He inspired many imitators, but they could never match the great man's talent and were quickly blown away. One lady, however, Angèle Thibeau, enjoyed a fair degree of success as a female Petomane.

She promised no trickery or offensive odor and even offered a money-back guarantee — customers only had to pay if they liked the show. Apparently, though, Madame Thibeau did resort to some chicanery, because she stopped performing when Pujol sued her, claiming that she used mechanical devices to produce sounds that to him came naturally.

Can we learn anything from Pujol's unique gift? He himself recognized the singular nature of his talent and accepted 25,000 francs from a medical school in exchange for his permission to examine his body after death. However, when the peerless performer passed on in 1892 at the age of 88, his children were not keen on pushing back the frontiers of science and nixed the postmortem. It is interesting to note that every morning the great Petomane cleansed his insides in his own singular fashion and was never sick a day in his life.

Some Final Views on the Nature of Science

I hope we've had some fun together and in the process shed some light on a few chemical mysteries. Perhaps, along the way, we've even learned some principles of chemistry and engaged in a little critical thinking. While I know that you'll probably forget most of what you read here, I hope the ideas I'm about to summarize will leave their mark. After all, it's been said that education is what you're left with after you've forgotten what you have learned.

1. Science is a truth-seeking process. It is not a collection of unassailable "truths"; it is, however, a self-correcting discipline. Such corrections may take a long time — the medical practice of bloodletting went on for centuries before its futility was realized — but as scientific knowledge accumulates, the chance of making substantial errors decreases.

2. Certainty is elusive in science and it is often hard to give categorical "Yes" or "No" answers to many scientific questions. To determine whether bottled water is preferable to tap water, for example, one would have to design a lifelong study of two large groups of people whose lifestyles were similar in all respects except for the type of water they consumed. This is virtually undoable. We therefore have to rely on less direct evidence in formulating many of our conclusions.

3. It may not be possible to predict all consequences of an action, no matter how much advance research has been done. When chlorofluorocarbons (CFCs) were introduced as refrigerants, no one could have predicted that 30 years later they would have an impact on the ozone layer. If something undesirable happens, it is not necessarily because someone has been negligent.

4. Any new finding should be examined with skepticism. A skeptic is not a person who is unwilling to believe; a skeptic is someone who bases his or her beliefs on scientific proof and does not swallow information uncritically.

5. No major lifestyle change should be made on the basis of any one study. Results should be independently confirmed by others. Keep in mind that science does not proceed by "miracle breakthroughs" or "giant leaps." It plods along, taking many small steps, slowly building towards a consensus.

6. Studies have to be carefully interpreted by experts in the field. An association of two variables does not necessarily imply cause and effect. As an extreme example, consider the strong association between breast cancer and the wearing of skirts: obviously, the wearing of skirts does not cause the disease. Scientists, however, sometimes show an amazing aptitude for coming up with inappropriate rationalizations for their pet theories.

7. Repeating a false notion often does not make it true. Many people are convinced that sugar causes hyperactivity in children — not because they have examined studies to this effect but because they have heard that it is so. In fact, a slate of studies has demonstrated that, if anything, sugar has a calming effect on children.

8. Nonsensical lingo can sound very scientific. An ad for a type of algae states that "the molecular structure of chlorophyll is almost the same as that of hemoglobin, which is responsible for carrying oxygen throughout the body. Oxygen is the prime nutrient and chlorophyll is the central molecule for increasing oxygen available to your system." This is nonsense: chlorophyll does not transport oxygen in the blood.

9. There will often be legitimate opposing views on scientific issues, but the notion that science cannot be trusted because for every study there is an equal and opposite study is incorrect. It is always important to take into account who carried out a given study, how well it was designed, and whether anyone stands to gain financially from the results. Be mindful of who the "they" is in "they say that . . . ." In many cases, what "they say" is only gossip, inaccurately reported.

10. Humans are biochemically unique. Not everyone exposed to a cold virus will develop a cold. Individual response to medications can be dramatically different. Eating fish can be healthy for many but deadly for those with an allergy (like me).

11. Animal studies are not necessarily relevant to humans, although they may provide much valuable information. Penicillin, for example, is safe for humans but toxic for guinea pigs. Rats do not require vitamin C as a dietary nutrient but humans do. Feeding high doses of a suspected toxin to test animals over the short term may not accurately reflect the effect on humans exposed to tiny doses over the longterm.

12. Whether a substance is a poison or a remedy is determined only by the dosage. It does not make sense to talk about the effects of certain substances on the body without talking about amounts. Licking an aspirin tablet will do nothing for a headache, but swallowing two tablets will make the headache go away. Swallowing a whole bottle of pills will make the patient go away.

13. "Chemical" is not a dirty word. Chemicals are the building blocks of our world — they are neither good nor bad. Nitroglycerin can alleviate the pain of angina or blow up a building. The choice is ours. Furthermore, there is no relation between the risk posed by a substance and the complexity of its name: dihydrogen monoxide is just water.

14. Nature is not benign. The deadliest toxins known, such as ricin from castor beans or botulin from the clostridium botulinum bacterium, are perfectly natural. "Natural" does not equal "safe," and "synthetic" does not equal "dangerous." The properties of any substance are determined by its molecular structure, not by whether it was synthesized by a chemist in a lab or by nature in a plant.

15. Perceived risks are often different from real risks. Food poisoning from microbial contamination is a far greater health risk than trace pesticide residues on fruits and vegetables.

16. The human body is incredibly complex and our health is determined by a large number of variables, which include genetics, our diet, our mother's diet during pregnancy, stress, level of exercise, exposure to microbes, exposure to occupational hazards, and pure luck.

17. While diet clearly plays a role in the promotion of good health, the effectiveness of specific foods or nutrients in the treatment of diseases is usually overstated. Individual foods are not good or bad, although overall diet may be described as such. The wider the variety of food consumed, the smaller the

chance that important nutrients will be lacking. There is universal agreement among scientists that a high consumption of fruits and vegetables is beneficial.

18. The mind-body connection is an extremely important one. The health of about 40 percent of test subjects will improve significantly when they are given a placebo, and about the same percentage will exhibit symptoms in response to a substance they perceive as dangerous. To paraphrase John Milton: the mind is capable of making a heaven out of hell and a hell out of heaven.

19. About 80 percent of all illnesses are self-limiting and will resolve in response to almost any kind of treatment. Often, a remedy will receive undeserved credit. Anecdotal evidence is unreliable, because positive results are much more likely to be reported than negative ones.

20. There is no goose that lays golden eggs. In other words, if something sounds too good to be true, it probably is. As H.L. Mencken once said, "Every complex problem has a solution that is simple, direct, plausible, and wrong."

21. Virtually any subject or issue that may arise will become more interesting and more complicated on deeper examination. Ours is a fascinating world.

22. Nobody has a monopoly on the truth, so don't guide your life by any set of rules produced by an individual. As Will Rogers declared, "Everybody is ignorant, only on different issues."

INDEX

Chemistry sets 50–53
Chicken feed 85
Chicken soup 118–21
Chinese Restaurant Syndrome 93, 96
Chloramine 229
Chlorine 17, 228–30, 232
Chlorofluorocarbons (CFCs) 62–65, 278
Chlorogenic acid 89, 110
Chlorophyll 279
Chocolate 130–32
Cholesterol 80, 88–90, 102–3, 124, 130, 133, 135, 138, 181, 194, 199
Cinchona tree 266
Citric acid 215
Clark, Barney 121
Cleopatra 178
Clostridium botulinum bacteria 99
Cobalt chloride 52–53
Cobaltous sulfate 53
Cocamide monoethanolamine 214
Cockroach 202–3
Cockroach perfume 203
Cocoa 130, 133
Coital frequency 248
Coital noise pollution 249
Cold 118–21, 242
Collagen 215
Collodion 41–42
Colon cancer 89, 196, 274
Colorectal cancer 191
Color Light Therapy Lamps 255
Color television 226
Color therapy 251–52
Colostrum 176–77
Conjurer 11
Constantine 98
Copper 46, 236
Copper sulfate 53
Cordus, Valerius 183
Cortes, Hernando 130

Cotton 40–41, 65–66
Coumadin 200
Coumaric acid 110
Counterfeiting 50
Counterfeit money 47
Courtaulds 67
Cow dung 268
Creosote 267–68
Crete 137–40
Crickets 15
Cronstedt, Baron A. F. 233
Crystaldyne pain reliever 238
Cuticle (hair) 215–16
Cutler, Winnifred 206–7
Cyanide 73, 157–58
Cyclosporin 76–77
Cysteine 118–19
Cytochrome oxidase 156

Daidzen,122
d-Alpha-tocopherol 197, 199
Datura stramonium 160
Davis, Jacob 46
Davis, Wade 159–61
Davy, Humphrey 183
Deadly Orgone Buster 259
Deadly Orgone Energy 259
Dead Sea 27
Defibrillation 235
Dégorgement 113
Demerol 155
Dentures 42
Deoxyribonucleic acid (DNA) 80, 108
Depression 115–16, 152, 154
Detergent 214, 219, 221–25, 234
DHA 116–18
Dichlorodifluoromethane 62
Dickson, William 39
Digitalis 195, 208–9
Dihydrogen monoxide 25, 280
Dihydrotestosterone 187–88

Kricket Krap 15, 22
Krieger, Dolores 256–57
Kwok, Dr. Ho Man 94

Lactose 273
l-Alpha-tocopherol 198
Lancet, The 60, 242
Laudanum 165
Laughing-gas (nitrous oxide) 183–84
Laundry Disks 221–23
Law of Infinitesimals 240
Lawrence, Joseph 58
Lead 179
Le Petomane 275–77
Licorice 171
Ligue des Femmes du Québec, La 63
Limburger 126–27, 129
Lime 29, 31–32, 175, 230, 271
Limelight 29–32
Limestone 30
Lind, James 175
Lister, Joseph 58–59
Listerine 58
Little Mermaid 224–27
Long, Crawford 185
Lot 27–29
Lot, Mrs. 27–29
Louis XV 111
Lovastatin 76
Love 130–31
Luftwaffe 82, 86
Lumière brothers 40
Lung cancer 83–84, 103
Lupercalia 98
Lwoff, Andre 120
Lycopene 110
Lyme disease 176
Lysergic acid diethylamide (LSD) 150

Ma Huang 155
MacMurray, Fred 21
Magendie 272
Magic chemical 11, 14
Magicians 69
Magic mushrooms 148
Magnesium 222–34
Magnesium bicarbonate 31
Magnesium carbonate 31
Magnesium hydroxide 246
Magnet 237
Magnus, Albertus 19
Maimonides, Moses 118
Malaria 240
Malic acid 75, 76
Manganese 179
Manganese dioxide 30
Mannitol 274
Marijuana 131
Match 262–65
Maynard, J. Parker 41
Medicomat 238
Memory 196
Mencken, H.L. 281
Meniere's disease 209–10
Meralgia paresthetica 45
Mercury 24, 34, 65, 225
Methane 230, 272–74
Methanethiol 273
Methemoglobin 158
Methionine 193
Methyl mercaptan 126, 234
Midgley, Thomas 62–63
Mikey 242–44, 246
Milk sugar 273
Miller's Tale, The 272
Millikan, Robert 238
Mind-body connection 281
Minoxidil (Rogaine) 188
Mitchell, William 243–44
Moët et Chandon 111
Moisture-wicking 66

Oxalic acid 70, 173
Oxazepam (Serax) 169
Oxidation 103, 199
Oxidizing agent 232
Oxygen 25, 30, 37, 58–59, 79–80, 156, 198, 221, 234, 251
Ozone 63–65, 80, 278

Pain 182
Panacea 194, 197
Panthenol 215
Paprika 99
Paracelsus 19, 163–66
Parkinson's disease 170
Parmesan 94, 96
Pasta 47
Pasteur, Louis 59, 67
Pava 167
Pectin 88–89, 134–36
Pekoe 102
Penicillin 279
Penicillum roquefortii 128
Pepper 99
Pepto-Bismol 171
Periplanone-B 202, 204
Perkin Medal 62
Perkin, William Henry 268
Pernod 210
Perrier 92
pH balanced 215
Phelan and Collander 42
Phenol 58–61
Phenylacetic acid 131
Phenylethylamine 131
Pheromone 201–8
Phosmet 87
Phosphates 223, 234. 262, 264
Phosphoglucomutase 44
Phosphorous 115, 262–65
Phosphorous oxide 263
Phosphorous, red 264
Phosphorous sesquisulfide 264–65

Phytoestrogens 123
Pig balls 57
Pigs 54, 57
Pine nuts 117
Ping-Pong balls 43
Piperine 118, 120
Piss prophets 173
Pittacal 268
Placebo 165
Plain and Easy Account of British Fungi, A 145
Plasticizer 41
Pliny 118
Pollen 177
Polyester 227
Polyethylene 54–58
Polygalacturonase 107, 109
Polymers 54
Polyphenols 102–4, 133
Polypropylene 65–66
Polysorbate 60, 189
Polyvinyl alcohol 87
Poppy 265
Pop Rocks 242–46
Porphyria 173
Potassium 118, 121
Potassium carbonate 157
Potassium chlorate 30, 263–65
Potassium cyanide 156
Potassium nitrate 37
Potato 47–49
Preti, George 206
Priestley, Joseph 182
Propecia 188
Propylene glycol 215
Prostate cancer 110, 124, 176, 191, 193
Prostate enlargement 187–88
Prostate-specific antigen (PSA), 110
Protein 80, 90–91, 93, 100
Psilocin 148
Psilocybe mushroom 148